Nov '95

Marcus →
With love,
Mum.

£1

Winds of Change

LIVING IN THE GLOBAL GREENHOUSE

John Gribbin and Mick Kelly

Headway • Hodder & Stoughton

For Chico Mendes, assassinated 22 December 1988, and the many others who have lost their lives fighting environmental destruction - and for those who continue to struggle.

This publication accompanies the television documentary 'Can Polar Bears Tread Water?' produced by Central Independant Television plc in association with Television for the Environment and The Better World Society.

ISBN 0 340 52283 6 boards
ISBN 0 340 51505 8 paperback

First published 1989

© 1989 John Gribbin and Mick Kelly

All rights reserved. No part of this publication may be reproduced or transmitted in any form or by any means, electronic or mechanical, including photocopy, recording, or any information storage and retrieval system, without permission in writing from the publisher or under licence from the Copyright Licensing Agency Limited. Further details of such licences (for reprographic reproduction) may be obtained from the Copyright Licensing Agency Limited, of 33–34 Alfred Place, London WC1E 7DP.

Typeset by Wearside Tradespools, Fulwell, Sunderland
Printed in Great Britain for the educational publishing division of Hodder and Stoughton Ltd, Mill Road, Dunton Green, Sevenoaks, Kent by Butler & Tanner Ltd, Frome and London

ACKNOWLEDGEMENTS

The authors and publishers would like to thank Central Television for making available material from interviews conducted during the making of 'Can Polar Bears tread Water?', and the World Wide Fund for Nature for their support.

Mick Kelly thanks Sarah Granich, Emma Kelly and Liz Kelly for inspiration, assistance and support. John Gribbin thanks Mary Gribbin for his political re-education, Jo and Ben for their constructive support during the writing of this book.

Both authors are grateful to their team at Hodder and Stoughton for the speed and precision with which their words have been brought to a wider audience.

Thanks are due to the following for the use of photographs:
Cover: Tony Stone Worldwide; Barnaby's Picture Library: 12, 27, 80(L), 90, 102(L), 113(L), 151, 152(B), 158; J. Allan Cash Ltd: 5, 13, 49, 57, 78, 84, 101, 115; Central T.V.: (iii), 33, 51(both), 56(all), 70(R), 73, 76, 80(R), 81(TR), 99(Both), 117(B), 118, 129(all), 131; L. Moore/Central T.V.: 22(L), 45, 89, 104(L), 137, 140, 159; Bruce Coleman Ltd: 2, 117(T), 119; Anne & Paul Ehrlich "Earth" 93; Farmers Weekly: 6(R), 52, 79, 81(L), 92, 94; Friends of the Earth: 153(L); Sally & Richard Greenhill Picture Library: 95(L); Hutchinson Library: 14, 82, 109, 153(R); National Trust/R. Judges: 91, 122, 123; North Norfolk District Council: 98; Oxfam: C. Flagg 18, O. Graham 68, J. Hartley 16(T), 23, 81(BR), 83, 97, 104(R), 106(B), 125(L), C. Williams 70(L); Panos Pictures: G. Bernard 106(T), 113 T. Bolstad 3,54, S. Cunningham 141, H. Giradet 15, P. Harrison 4,96, B. Press 40, D. Reed 105, M. Santilli 60, A Van Buren 108; Population Concern: 28; Royal Society for the Protection of Birds: 74; J. Sainsbury Plc: 154; Science Photo Library: M. Bond 1, 72, Dr. J. Burgess 7, S. Fraser 47, M. Gilbert 11, P. Jude 30, Prof. S. Lowther 38, NASA 31, 32, 36(Both), D. Parker 39, C. Raymond 95; Tony Stone Worldwide: 86; Survival International/Steve Cool: 145; Topham Picture Library: 22(R), 88, 136; Toyota (GB) Ltd: 102(R); WWF-UK: B. Chapman 16(B), B. Massey 143, M. Payne-Gill 20, J. Plant 152(R).

This book has been printed on non-chlorine bleached paper, produced in Sweden. Independent tests by the Swedish Environmental Research Group confirm that the paper mill concerned, Papyrus Nymölla AB, is the first and so far the only pulp mill producing bleached pulp in which dioxin contaminants do not occur.

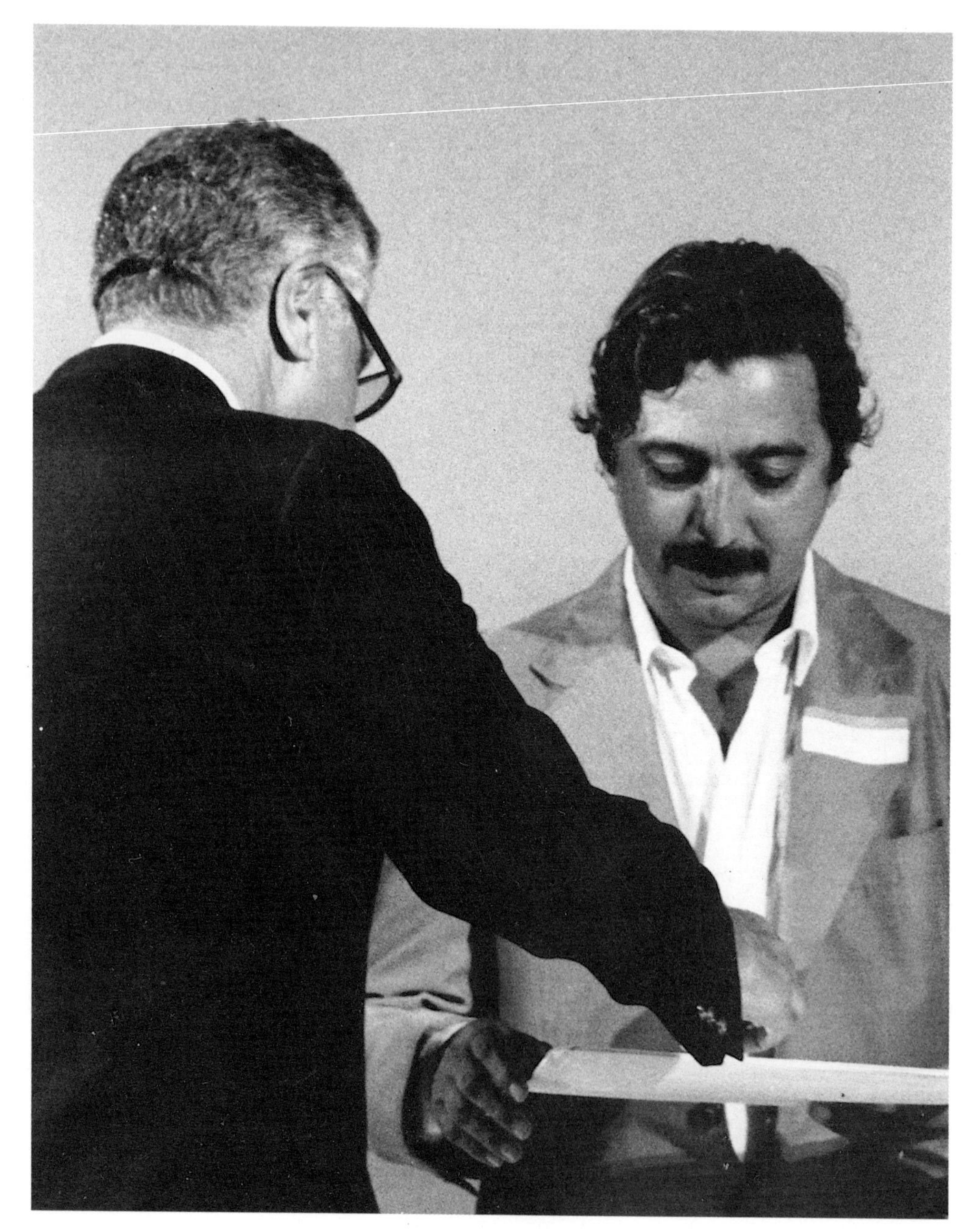

Chico Mendes receiving the UNEP Global 500 Award in 1987

CONTENTS

1

THE ROAD TO RUIN

June 1988. As the North American continent basked in the early days of one of the driest summers on record, 300 scientists, environmentalists, decision-makers and politicians met in Toronto, Canada, to discuss the threat of environmental catastrophe. In the terse words of the conference statement: 'Far-reaching impacts will be caused by global warming and sea level rise. . . . The best predictions available indicate potentially severe economic and social dislocation for present and future generations, which will worsen international tensions and increase the risk of conflicts among and within nations. It is imperative to act now.'

The world's weather is changing.

The average temperature of our planet has increased by about half a degree Celsius (°C) since the middle of the nineteenth century. This may not seem much compared to the swings in temperature that occur from day to day but it represents a major change in global climate. The decade of the 1980s has been the warmest since reliable observations began, with records broken almost every year. The unprecedented warmth of 1980 was surpassed in 1981. That record fell in 1983. 1987 set a new all-time high, only to be beaten in 1988. Even the coldest year of the 1980s – 1985 – was warmer than any year before 1930. As temperatures have risen, so has sea level. Since the turn of the century, global sea level has risen by approaching 15 cm. Almost all of this can be explained by the expansion of sea water as the oceans have warmed. But some of the rise is due to the melting of mountain glaciers, on the retreat in many parts of the world. As glaciers melt, the water runs off – eventually reaching the sea.

For many people, water availability is the critical factor determining either survival or death. Recent analyses have revealed startling trends, signs of a global shift in rainfall patterns. Whilst rain and snowfall levels over middle to high northern latitudes have been increasing in recent decades, rainfall over the subtropics has been in decline. The drought in northern Africa that has underlain the tragic famine of recent years is the most pronounced manifestation of the subtropical trend, but other areas, such as the Caribbean and southeast Asia, have also been affected.

'It's time to stop waffling so much and say the evidence is pretty strong that the greenhouse effect is here.'
Jim Hansen, NASA

Every year brings news of climatic disasters in one part of the world or another. As Lester Brown and his colleagues at the Worldwatch Institute at Washington DC have observed, as concern about global warming mounts weather conditions around the world seem 'to preview what life in a greenhouse world would be like.' July 1988. Central China experienced an unprecedented heat wave. Temperatures reached 40°C and, as hospitals were overwhelmed by the victims of heat stroke, hundreds died. Autumn 1988. Isolated by floods, 8000 people are reported to have starved to death in the southwest of Sudan. Torrential rains and flooding devastate Bangladesh, sub-

merging two-thirds of the country and leaving 20 million people homeless. In North America, continuing drought means that, for the first time in many years, the United States – the breadbasket of the world – consumes more grain than it produces.

That same season, one of the most powerful hurricanes of the twentieth century, Hurricane Gilbert, scythed through the Caribbean. Sweeping in a 2500 mile arc from St Lucia in the eastern Caribbean to the Yucatan Peninsula of Mexico, it killed hundreds, left hundreds of thousands homeless and caused billions of dollars worth of damage. Just one month later, Hurricane Joan grazed the northern coast of South America and cut a swathe through Central America. The strongest hurricane ever to hit Nicaragua, Joan unleashed 170 mile an hour winds on the Atlantic coast, destroying towns, roads, bridges, crops, wildlife and forest.

We are seeing the first signs of a change in climate without parallel in human history. Pollution shrouds the planet, trapping heat near the Earth's surface and preventing it from escaping to space. The process causing the heat trap is known as the 'greenhouse effect' and the gases responsible are known as 'greenhouse gases'.

> 'Time is passing quickly. Atmospheric concentrations of greenhouse gases increase steadily. Significant climate change already may be unavoidable given current and historical rates of emissions. If governments conclude that a substantial further warming poses unacceptable social risks, then strong actions need to be taken, and taken soon. The longer the delay before preventive policies are identified, agreed upon, and implemented, the more extreme the policies imposed to stay within prudent bounds will be.'
>
> Irving Mintzer, World Resources Institute

With global temperature at a record high, any further warming will move the climate system into uncharted territory. The past can no longer provide a reliable guide to the future. Farmers in the midwest of the US, nomads in the Sahel, rice growers in Japan, the industrialists of western Europe, we may all have to learn to live in a very different world than that to which we are accustomed. As Gro Harlem Brundtland, Prime Minister of Norway and Chair of the World Commission on Environment and Development, has warned: 'We have come to a threshold. If we cross this threshold, we may not be able to return.'

INTO THE HEAT TRAP

Global warming results from the rise in energy consumption, the expansion of industry and the intensification of agriculture that has occurred since the industrial revolution. It is the most striking indication of the pronounced transformation in the relationship between humanity and the natural environment that has occurred over recent centuries. Whether it be through erosion of the planet's soils, institutional policies which lead to environmental degradation or pollution of the atmosphere, we are now capable of changing the planetary environment on a scale that is without precedent. We are moving into the heat trap.

Before the invention of agriculture, the world population remained stable at around 5 million people. As hunting and gathering gave way to farming, population levels rose – to around 250 million at the time of the birth of Christ, 500 million by the Middle Ages. The current phase of explosive growth began at the time of the industrial revolution, in the eighteenth century. The world population reached 1000 million some time around 1800, 2000 million in 1930, 3000 million in 1960, 4000 million in 1975 and 5000 million in 1987. The need to feed an ever-increasing number of mouths is a driving force behind the intensification of agriculture. Worldwide, agricultural production has been growing at about two and a half percent per year during recent decades – it doubled between 1950 and 1980. This is a faster growth rate than that of population during that period. No-one would

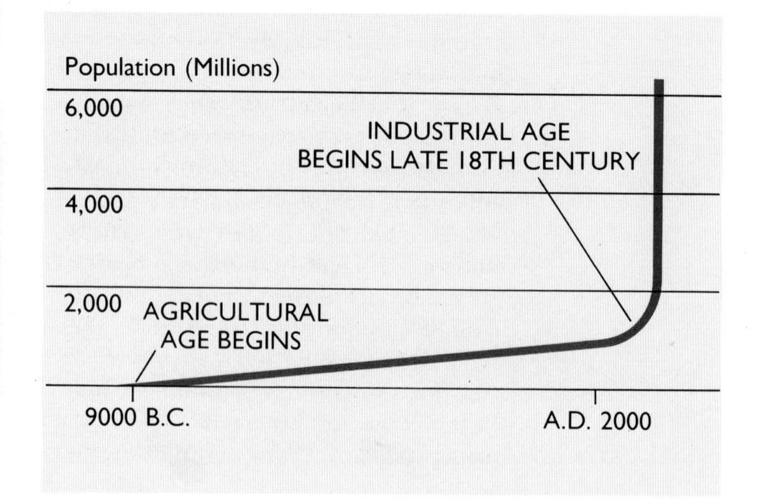

World population 9000 BC – AD 2000 Source: World Bank/The Hunger Project

argue that improving agricultural productivity is a bad thing but, as we shall see, agricultural activity is a principal source of the greenhouse gases.

Greenhouse gas: *carbon dioxide*

Trees, and other forms of plant life (including the microscopic plankton that live in profusion in the surface layers of the sea) regulate the amount of carbon dioxide in the atmosphere. They are the lungs of the planet. Plants take in carbon dioxide and water and, with the aid of sunlight, use the carbon to build up their tissues as they grow – a process known as photosynthesis – while releasing oxygen back into the air. When plants die and decompose, or if they are burnt, the carbon they contain combines with oxygen once again to make carbon dioxide.

Carbon is therefore constantly being cycled through the atmosphere and biosphere (the world's living matter). But measurements of the amount of carbon trapped in tiny bubbles of air in ice cores drilled from the polar glaciers show that its atmospheric concentration has scarcely varied from 280 parts per million for more than 10,000 years, since the end of the most recent ice age. The natural carbon cycle kept the proportion of carbon dioxide in the atmosphere constant – until humanity interfered. The destruction of the world's forests releases about 2 billion tonnes of carbon dioxide directly into the air each year as the trees and other forest material are burnt or decay. It also reduces the ability of the biosphere to absorb the carbon dioxide being released by other carbon dioxide producing human activities, such as burning fossil fuel.

There are many reasons why forests are being cleared but a major one is to release land for agriculture. This is not simply to feed local people but to provide cash crops for export or beef for the fast-food industry of the North. 'More than 40 percent of the Brazilian forest lost in recent years has been due to clearing for cattle production, mainly for export to US fast food outlets,' says Ted Trainer, author of *Developed to Death*.

As well as contributing to the buildup of carbon dioxide in the atmosphere, switching land over to agriculture reinforces the heat trap by releasing other greenhouse gases such as methane.

Greenhouse gases: *methane and nitrous oxide*

Like carbon dioxide, methane is a natural part of the atmosphere, a gas that cycles through living systems of the biosphere. It is produced by bacteria that live in places where there is very little oxygen, such as swamps and marshes: 'swamp gas' is its common name. Rice paddies, in effect artificial swamps, are a major source of methane; it is also produced by the bacteria that live in the guts of cows and other ruminants (and in termite mounds). Bizarre though it may seem, cattle farming is a significant contributor to the greenhouse effect, causing the release of about 90 million tonnes of methane to the atmosphere each year in the form of cow belches. Rice paddies produce even more methane, about 120 million tonnes a year. Natural gas escaping from coal mining and methane released when forests are burnt, also add their burden to the atmosphere, along with the natural production of the gas from swamps and termite mounds.

The buildup of methane is particularly alarming because each molecule of methane is 20 times more efficient at trapping heat than a single molecule of carbon dioxide. Whereas the amount of carbon dioxide in the air has increased by a quarter as a result of human activity, the amount of methane has doubled over the past 200 years, and is now increasing at a rate of one percent a year – twice as fast as the buildup of carbon dioxide. It is hard to imagine how this can be stemmed in a world where more food is needed each year to feed an ever-increasing population. Arable farming also reinforces the heat trap by releasing another greenhouse gas, nitrous oxide. This is produced from the nitrogen fertilizers that are added to fields in such large quantities. It is also a byproduct of combustion – when anything burns in air, which is four-fifths nitrogen, some nitrous oxide is produced as a result.

As a source of greenhouse gases, agricultural activity is only surpassed by our demand for better standards of living, improved lifestyles. This demand is being met by the expansion of industry and increasing energy consumption. Industrial pollution and the burning of fossil fuels – coal, oil and gas – to generate energy produces a variety of greenhouse gases, of which carbon dioxide is the most important. In the Third World, the burning of biomass, such as fuelwood, is also a major source of carbon dioxide. While the world population has tripled since 1900, global energy consumption has increased by a staggering factor of ten. As much coal has been burnt in the past four decades as had been consumed in the whole of previous human history. If all the extra energy were distributed evenly around the world, that would mean that each person was using more than three times as much as their counterparts at the turn of the century. That should translate, roughly, into a three-fold improvement in the standard of living for every person on the planet.

False-colour photograph of Ferrybridge coal-fired power station in Yorkshire, England, highlighting the pollutants emitted from the smokestacks

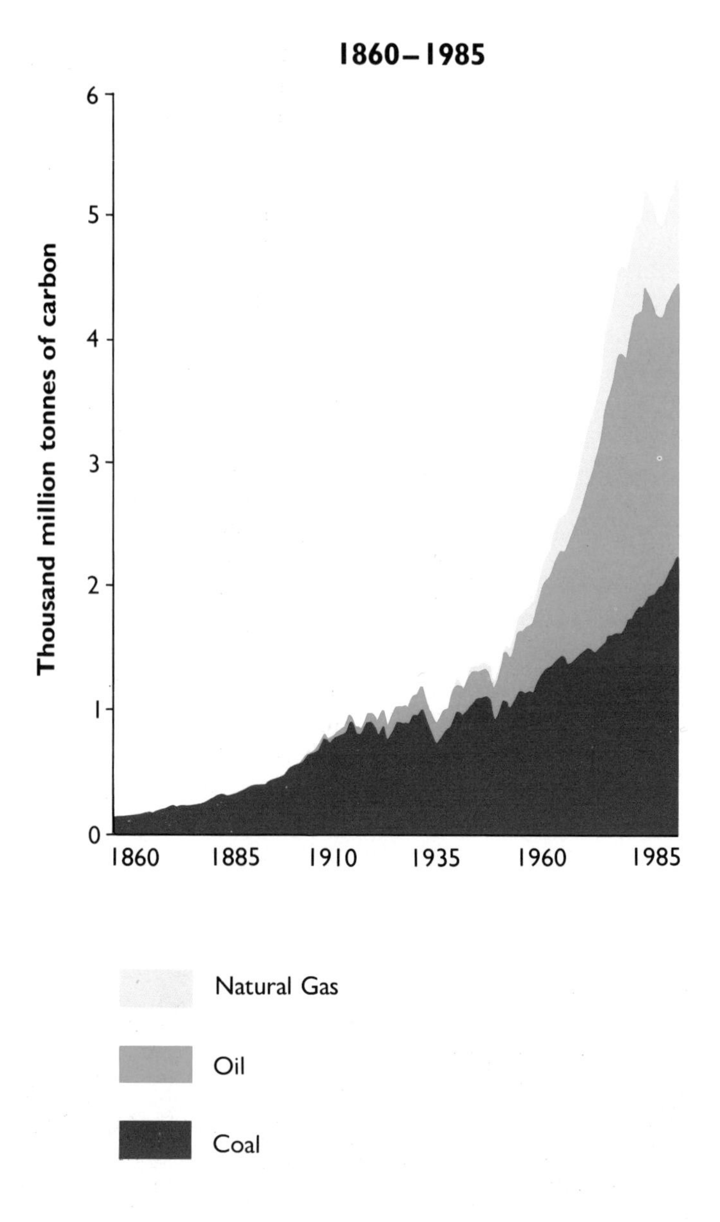

Carbon dioxide emissions due to fossil fuel combustion
Source: R M Rotty and C D Masters

> 'After 300 years of virile growth, and a long boom in which doubt was inconceivable, the growth and greed society has suddenly burst into an area of problems, chaos and uncertainty, giving us an unprecedented opportunity to move to more sensible arrangements. Whether or not we manage to take that opportunity will depend on whether enough of us devote ourselves to the crucial educational task. Unless many more of us do adopt this as own chief long-term priority, there is little chance of us making it to a just, peaceful and ecologically sustainable world order.'
>
> Ted Trainer, ***Developed to Death***

In fact, most of the additional energy is consumed in the rich nations of the industrialized world, usually referred to as the 'North', but including Australia and New Zealand. This energy is used, for example, to heat your home in winter and keep it cool in summer. Your car uses energy when it burns petrol. But this additional energy consumption also takes place in the factory where your car was built, in the process of making the bricks to build your home, in manufacturing the clothes on your back and on the farms that produce the food that you eat.

Most of the increasing population of the globe still has only a very limited access to sources of energy, and to the benefits that the use of energy has brought the industrialized world. Three-quarters of the world's energy is consumed by one quarter of the world's population. An inhabitant of the Third World consumes, on average, a sixth of the energy used by somebody living in an industrialized nation. The average American uses 400 times as much energy as the average Ethiopian. Even if the population of the world remained constant over the next 100 years, the efforts of the developing countries to match the development of the industrialized world would still ensure a continuing increase in the global consumption of energy, and all the problems that that will bring.

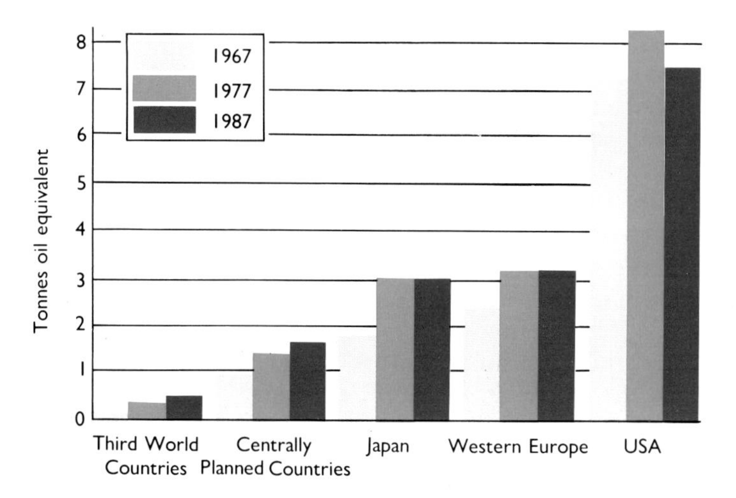

Energy consumption per head of population Source: British Petroleum

Most energy – well over 90 percent – is produced by burning fossil fuels such as coal, gas and oil, or wood, all of which contain carbon. Energy is released when the carbon in the fuel burns in air to produce carbon dioxide, and carbon dioxide is the main contributor to the greenhouse effect. The world burns so much fuel that we release 5.5 billion – 5,500,000,000 – tonnes of carbon into the air each year. Like all such astronomical numbers, this doesn't mean much more than a lot of zeroes. But we can put it into perspective by comparing the amount of carbon dioxide in the atmosphere now with what was there before the industrial revolution.

In the nineteenth century, the proportion of carbon dioxide in the atmosphere was about 280 parts per million, or 0.028 percent. This is a very small proportion of the whole atmosphere but important nevertheless. By the end of the 1980s, the amount of carbon dioxide in the air had increased by a quarter, to 350 parts per million. Almost half of this increase has occurred since 1960 as the pace of global development has accelerated. At present, the carbon dioxide concentration is increasing by about one and three-quarters parts per million each year, a growth rate of 0.5 percent a year.

Today, most of this extra carbon dioxide comes from our burning of fossil fuel although half of the buildup over historical times has been caused by

Carbon dioxide concentration in the atmosphere measured in parts per million by volume

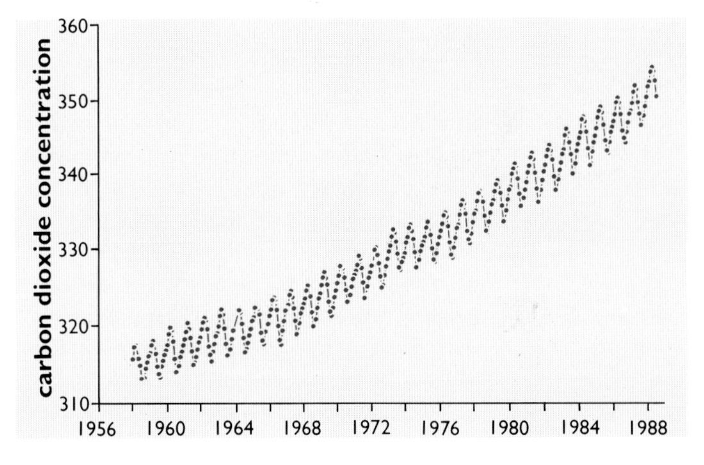

the destruction of the world's forests and clearance of new land for agriculture.

All of the greenhouse gases we have discussed so far – carbon dioxide, methane and nitrous oxide – are natural components of the atmosphere. It is simply their quantities that have been changed by human activity. But the last major greenhouse gases contributing to the heat trap have no parallel in nature. They are artificial products of our technology. The chlorofluorocarbons, or CFCs, never existed in the atmosphere at all until manufactured by industrial chemists in the 1930s.

Greenhouse gases: *chlorofluorocarbons*

CFCs are now notorious as the gases that lead to destruction of the ozone layer in the upper atmosphere, tens of kilometres above our heads; but they are also extremely efficient greenhouse gases. Molecule for molecule, some CFCs are 10,000 times more effective at trapping heat than carbon dioxide. Fortunately, as yet they are present in the air only in tiny quantities. But, even with all the concern about damage to the ozone layer, the two main CFCs are still building up in the atmosphere at a rate of about six percent a year – and because there are few natural mechanisms for their removal, they stay in the atmosphere for so long that only about ten percent of all the CFCs ever released have yet been broken down and decomposed into other substances.

When they were first invented, in the 1930s, CFCs seemed like miracle chemicals. They are non-toxic and non-inflammable, and have almost ideal properties for use as the working fluids in refrigerators, as the gases that make the spray in spray cans, and as insulators to make the bubbles in plastic foams. They can also be released safely in close proximity to human beings. They soon found a wide variety of uses in the industrialized world, and the amount of CFCs being released into the air increased rapidly into the 1980s. It was then realized that, although CFCs may be safe in the short-term, there were long-term consequences which had not been recognized.

The problem with CFCs is that they have a long lifetime once released into the air; they only break down when they reach the upper atmosphere, releasing chlorine. This attacks ozone and thereby allows more solar ultraviolet radiation, usually blocked by the ozone shield, to penetrate to the surface of the Earth. So far, ozone depletion is particularly pronounced over Antarctica where a 'hole' in the ozone layer appears each spring. The depletion of ozone occurs strongly over the south polar region because the air there is very cold and still in winter, and this allows chemical reactions to take place that occur nowhere else on Earth. But the air over the arctic region is only slightly less cold and almost equally still in wintertime,

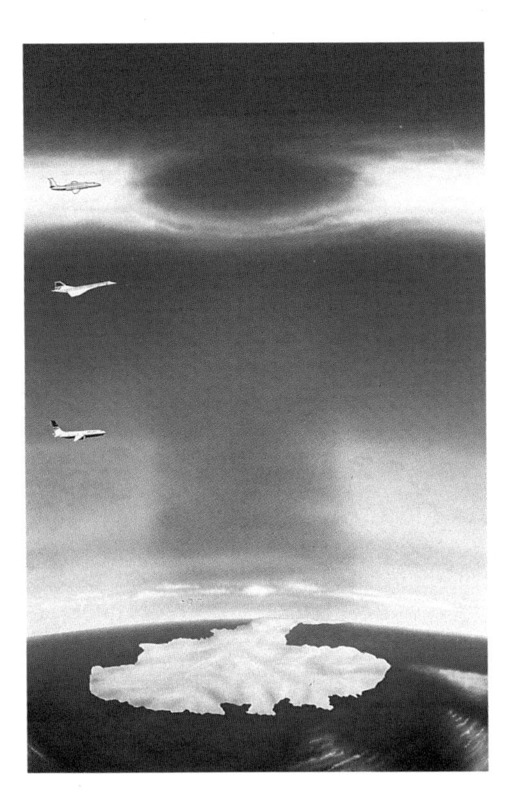

Representation of the ozone hole high over Antarctica

and there are signs that similar chemical reactions are beginning to take place there. This could mark the first stage of the spread of the influence of CFCs over populated regions of the Northern Hemisphere.

The amount of ozone in the stratosphere between 53°N and 64°N declined by six percent between 1969 and 1986. Over the same period, the decline between 40°N and 52°N was just under five percent. These latitude bands encompass all of Europe north of the Mediterranean and Black Seas, and North America north of a line from San Francisco to Washington DC. Measurements made from a high-flying aircraft which flew into the region of polar ozone depletion have established that the main cause of this ozone decline is the effect of chlorine atoms released by the breakdown of CFCs in the stratosphere – but nitrous oxide also contributes to the breakdown of ozone, in a more modest way, forging yet another link between global pollution problems.

Ozone depletion presents a serious threat. A ten percent decrease in the amount of ozone in the stratosphere would cause a 20 percent increase in the amount of damaging ultraviolet radiation reaching the surface of the Earth. This in turn could cause a rise of 40 percent in the number of non-lethal skin cancers among fair-skinned people, as well as a lesser increase in deaths from a malignant, but more rare, form of skin cancer. But the ozone problem is not something that dark-skinned people can ignore. An increase in exposure to ultraviolet causes a general depression of the human immune system, making the body less resistant to disease. It also provokes eye cataracts – one of the main causes of blindness in the Third World – and damages both livestock and crops in the fields, as well as plankton, fish larvae and other creatures that live in the surface layers of the sea.

According to Bob Watson of NASA, a world authority on ozone depletion, the Antarctic ozone hole would be with us for at least another century even if we took dramatic action immediately. The lifetime of the CFCs in the air is so long that, to hold the concentration of these gases steady at the level which has already begun to damage the ozone layer, releases would have to be cut by 85 percent. For all practical purposes, a total ban on CFCs is required to give the ozone layer a chance to begin to recover from the damage that has already been done. And therein lies another problem.

Chemists trying to find alternative gases that can take over the job of CFCs in, for example, spray cans have concentrated on manufacturing products that do not harm the ozone layer. All well and good, but it turns out that some of the alternatives now coming into use are, like the CFCs they are replacing, very efficient greenhouse gases. There is little point in 'solving' the ozone problem if we strengthen the heat trap at the same time. All of these problems of environmental pollution have to be looked at together, in a global context, before appropriate steps to minimize the impact of human activities on the environment can be decided upon. By the same token, if we can slow the buildup of greenhouse gases we will go a long way towards solving a range of other problems as well.

THE DEVIL'S COCKTAIL

Global warming is just one of the many inter-related hazards – such as photochemical smog, acid rain and depletion of the ozone layer – caused by air pollution. These problems are linked both by source, because the same chemicals released into the atmosphere contribute to more than one problem, and by impact. For example, ecosystems and crops stressed by acid rain and ozone depletion will be more vulnerable to climatic change. Forest loss caused by acid rain contributes to the greenhouse effect.

Greenhouse gas: *ozone*

The links are clear in the case of ozone. Ozone in the upper atmosphere, the stratosphere, is a good thing, shielding us from harmful ultraviolet radiation from the Sun. But ozone near the ground is highly undesirable. It contributes to photochemical smog, to acid rain, and is also a greenhouse gas. Near the Earth's surface, ozone is building up over the northern continents at a rate of about two percent a year.

Ozone, a major constituent of photochemical smog, is produced by the action of sunlight on nitrogen oxides, hydrocarbons and carbon monoxide – emitted, for example, from vehicle exhausts. It forms where the air is still and there is plenty of sunlight, spreading its pall over the cities which are the source of the pollution and over the surrounding countryside. It was first evident in large urban areas in the industrialized world – the classic example is Los Angeles – but is now becoming a common feature in many parts of the Third World.

Although air pollution is becoming a widespread problem in the Third World, there are few laws to limit emissions and control measures are rarely implemented. The situation is made more serious because the polluting industries are now restricted in the industrialized world by environmental regulations. It pays the companies that own the factories to switch production to the Third World where there is no such legislation rather than to meet the costs of environmental control at home. The results can be seen in the 'Valley of Death' – Cubatão – in Brazil where

uncontrolled industrial development has created an ecological wasteland and resulted in the remoseless degradation of human health. Living in the most polluted city in the world, the population of Cubatão suffer horrendous rates of infant mortality, birth defects, lung diseases and skin disorders which would not be tolerated in the North. Transnational companies are guilty of blatant hypocrisy when they follow one standard in the North and another, lower, one elsewhere.

Cubatão

Ozone is a poison, affecting plants, animals and people. In some parts of the world, the level of ozone in the air near the ground is so high as to be near the limit that plants can tolerate. Where there are large urban sprawls, such as in central Europe and the US, ozone levels are high even outside the city boundaries. This may be an important factor contributing to the declining health of forests in parts of Europe and North America. The San Bernadino forest east of Los Angeles is thought to be suffering significant ozone damage. Apart from trees, many crops, such as potatoes, grapes and dwarf beans, are at risk from increased concentrations of ozone.

Ozone is also implicated in the acid rain problem, compounding damage caused by other pollutants such as the nitrogen and sulphur oxides. It reacts with other pollutants to form acid in the air. Nitrogen oxides and sulphur oxides from vehicle exhausts and power plants increase the acidity of rainfall and of the environment at large. According to a recent United Nations report, eight European countries are rated as suffering 'severe' forest damage due to acid rain and ozone – West Germany, Liechtenstein, Switzerland, the Netherlands, Denmark, Britain, the USSR and Czechoslovakia. In western Europe, an area of forest the size of West Germany is at risk. In West Germany itself, the Black Forest is losing about one third of its trees. This represents a loss of some $800 million each year to the timber industry. Agriculture faces losses of around $600 million a year because of effects on soil fertility. Don Hinrichson, former editor of the *World Resources Report*, notes that 'damage to metals, building exteriors and painted surfaces alone [has] cost the 24 member countries of the OECD some $20 million a year.' The problem is more extreme to the east where human health is directly at risk: in Czechoslovakia, water supplies from shallow wells have become so polluted that pregnant women and young children are officially advised to drink bottled water.

'Acid rain spares nothing. What has taken humankind decades to build and millennia to evolve is being impoverished and destroyed in a matter of a few years – a mere blink in geologic time.'
Don Hinrichson, *The Earth Report* (Mitchell Beazley, 1988)

Crossing national boundaries, acid rain not only affects crops and ecosystems but can heighten tension between countries. About half of West Germany's acid rain problem is imported from other countries. The US exports acid rain into Canada but imports pollution from Mexico. In South Korea, the important rice-growing region of the Sasmoon Plains is badly affected by acid rain originating in mainland China. Negotiations have been underway in North America and Europe for many years to control the scale of transboundary pollution – but with only limited success. George Stewart, Professor of Botany at University College, London, greeted UK government plans to build the world's largest sulphur removal plant to meet European Community anti-pollution requirements with the wry comment: 'It's welcome, but it's a bit like closing the stable door after the horse has bolted.'

He reckons that some areas of Wales and southern Scotland could take many years to recover.

Regarded as a limitless resource, our planet's atmosphere has long been used as a dumping ground for unwanted chemicals. If local pollution proved a problem, higher chimneys would be built to spread the pollution further afield so that it became someone else's problem. Out of sight, out of mind – until the neighbours complained. It was assumed that in time the atmosphere would cleanse itself. We now know that this is not true. As we have seen, pollutants can stay in the atmosphere for decades or even centuries. The net effect of our mindless pollution of the air has been to produce a substantial change in the composition of the global atmosphere – and only now are we beginning to realize the global nature of the consequences.

THE EARTH STRIPPED BARE

The fall of the rainforests to cattle and arable farming directly affects the lives of many inhabitants of the Third World, but the destruction has implications for us all. 'What we see in Amazonia (and in all other rainforests of the world) is perhaps the most blatant example of how modern industrialized society has gone wrong, of how it has embarked on a suicidal course,' warns Jose Lutzenberger, the Brazilian ecologist.

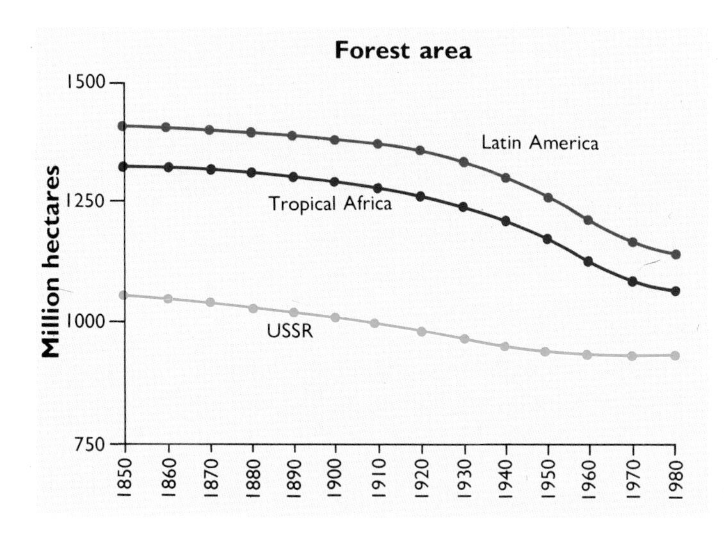

Rate of deforestation 1850-1980

The global dimension is also clear when considering the fate of the world's forests. An area of tropical forest equivalent to the size of the UK is vanishing every year – over 1 million acres a week, over 100 acres a minute. The rainforest ecosystem is one of delicate balance, sustaining life for every creature on Earth. The habitat of this vital ecosystem is being destroyed at such a rate that it may well cease to exist within the next 50 years. As well as enhancing the greenhouse effect, as we have seen, through the release of carbon dioxide and other gases, deforestation means that a major resource is being lost, at a fraction of its true worth.

'In 1950 15% of the Earth's area was covered in tropical forests. By 1975 it had declined to 12% and, at the present rate of elimination, by the year 2000 we could well have less than 7% of tropical forests remaining.'
Norman Myers, *Gaia: An Atlas of Planetary Management*

The rainforest provides humanity with an irreplaceable range of services. Services that include regulation of the composition of the atmosphere, amelioration of the local climate, provision of fresh water, disposal of wastes, recycling of nutrients and maintenance of a genetic reservoir of unparalleled diversity. Tropical forests, some of which have evolved over 180 million years, can give life to over 100 species of animal in less than two and a half acres. Of the world's primates, 90 percent are found only in tropical rainforests, along with two-thirds of all known plants, 40 percent of birds of prey and 80 percent of the world's insects. As a result of tropical deforestation at least one species is condemned to extinction every day. Pharmaceutical companies have drawn heavily on the genetic reservoir of the tropical forests – over a third of all medicines have their origin in the rainforest. At the present rate of deforestation, over 2000 species of plants, many with anti-cancer properties, could be lost by the year 2000.

Medicines, food and industrial products from the rainforests

- 70% of the 3000 plants identified by the US National Cancer Institute as having anti-cancer properties come from the rainforests.
- Alkaloids yielded by the Madagascan forest plant, the rosy periwinkle, have increased the recovery rate from childhood leukaemia and other blood cancers from 20% to 92%. Commercial sales of the drugs in the United States are worth $175 million a year.

- An average of one in four of all purchases from Western high street chemists contain compounds derived from rainforest species.
- At least 1650 known tropical forest plants have potential as vegetable crops.
- The berry of the serndipity fruit from West Africa's forests is 3000 times sweeter than sucrose.
- 50,000 tons of nuts from the Brazil nut tree, *Bertholettia excelsa*, are harvested annually. Apart from being a well known food source, oil extracted from the nuts is used for cooking, lighting and soap. Because of the complex ecology of the tree, all attempts to grow it in plantations have so far failed.
- The sap of the Amazonian tree *Copaiba langsdorfia* is so similar to diesel fuel that it can be put straight into truck engines.
- The value of the world trade in rattan exceeds $1 billion annually.

Nowhere is the complex mesh of ecological balances more clearly under threat from human activity than in the case of the tropical forests. The forest ecosystem is rich but the soil is fragile, dependent as it is on the nutrients absorbed by the diversity of plant life. Once that plant life is removed, a minimal soil quality is left for farming. Where once the moisture filtering through the forest canopy cycled nutrients through the ecosystem before replenishing the fresh water supply, now the rain rushes down the bared earth snatching away the topsoil. Streams and rivers are flooded with silt and debris, leaving a mere five percent of the water to the futile task of nourishing a plundered soil. With the loss of natural constant recycling of rainwater through the forest and back into the air, the supply of moisture to surrounding areas dries up. As the cloud cover of moisture is reduced and the soil-protecting natural vegetation replaced by crops, the albedo – the extent to which solar energy is reflected – is altered. The inevitable result is increased climatic disruption.

And, again, global warming is enhanced. As the complex balance of the rainforest ecosystem is destroyed, deforestation sets in motion a chain of processes which amplifies the effects of the initial release of greenhouse gas carbon dioxide into the atmosphere: the denuded soil left by forests stripped for crop-farming or cattle-ranching is only productive if fed with massive amounts of fertilizer from which greenhouse gas nitrous oxide is released; even with fertilizer, the soil can only be commercially productive for a mere four or five years

after which nothing remains but a barren landscape; finally, a third greenhouse gas, methane, is produced as organic carbon compounds break down in water-logged soils and in the digestive systems of the cattle reared to provide cheap beef for the fast-food industry.

Edward Goldsmith, Nicholas Hildyard and Peter Bunyard, editors of the *Ecologist*, put it succinctly: 'Clearly, a radical new approach is required if deforestation is to be halted and a global catastrophe is to be averted. The forests cannot possibly be saved if we continue to see them as but another resource to be cashed in. They are indeed a resource, but not because they can be transferred into commodities to be sold on the open market. They are a resource in the sense in which the planet itself, the sun and the atmosphere are resources; they make life possible and must therefore be preserved in a state which enables them to do so.'

A CRISIS OF OUR OWN MAKING

Global air pollution presents a serious threat to the welfare of humanity, 'second only to nuclear war' in the words of the participants of the 1988 Toronto Conference. Why are we faced with environmental catastrophe? As we have seen, the processes that are giving rise to this unprecedented change in atmospheric composition – the increase in energy consumption, the intensification of industry and the expansion of agriculture to meet the rise in population levels and the demand for a better quality of life – are fundamental to the development of modern civilization. Is it, perhaps, the model of development that we have followed that is to blame for the problem? No-one would argue that feeding the world and striving for a better standard of living is wrong; but maybe we have gone about achieving these aims in the wrong way.

Consider *energy consumption*

One of the strategies proposed at the 1988 Toronto Conference in order to moderate the greenhouse effect was to limit carbon dioxide emissions to 80 percent of the present-day value by the year 2005. How? By reducing wasteful energy consumption. This could be done. Using energy saving technology already available, such as the energy-efficient light bulbs illustrated, Jose Goldemberg and his colleagues at the World Resources Institute (WRI) in Washington DC have calculated that the energy used by each person in the industrialized world could be cut by half over coming decades without penalty to our lifestyle. This would have a range of benefits, whether or not the world warms substantially. A world adopting an energy-saving strategy would, the analysts note, be 'more equitable, economically viable and environmentally sound. It would also be more conducive to achieving self reliance and peace. And, it would offer hope for the long term future.'

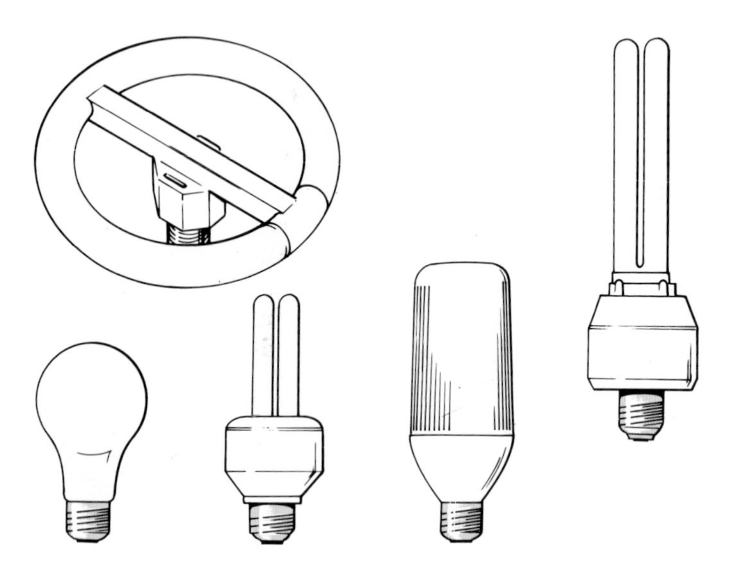

Energy-efficient light bulbs
Source: World Resources Institute

In recent years manufacturers have been steadily improving the compact fluorescent light bulbs that can be screwed into ordinary incandescent sockets. These bulbs have efficacies in the range of 40–60 lumens per watt, compared to 11–16 lumens per watt for ordinary incandescent bulbs. Although the first generation 'circline' bulb (above on the left) is efficient, it is bulky (it weighs 1 lb [454 g] and is 8 ins [20 cm] across) and thus does not fit easily into many lamp holders. The harsh white light of the circline is also considered inappropriate for many household applications. Finally, there is a lag after the switch is flicked before the light comes on – an annoying feature for those accustomed to the instant-on feature of incandescents. Each of the three bulbs to the right of the circline incorporates innovations that deal with one or more of these problems. The bulb shown next to the incandescent on the right, manufactured by Osram and now being sold in Europe, has none of the drawbacks of the earlier bulbs. It provides as much light as a 60-watt incandescent drawing only 11 watts and lasts six times as long. It weighs just 4 oz [114 g] and fits into most incandescent lamp holders. The phosphor coatings on the inside of the bulb give rise to a light quality similar to that for incandescents. For the impatient, the light is instant-on.

Yet we persist in being extremely wasteful and inefficient in our use of resources.

Consider *daily household waste*

The level of waste in the industrialized world is atrocious. Billions of pounds are thrown away each year due to the inefficient use of paper, plastics and metals. According to Norman Myers, editor of the *Gaia Atlas of Planetary Management*, only a quarter of the world's paper is recycled even though it would be feasible to double this figure within ten years or so. This would preserve 8 million hectares of forest and it would also save the energy needed to process the raw material. The Body Shop, marketing ecologically-sound cosmetics,

uses recycled paper bags in its outlets. The bags tell a home truth: that recycling paper could save the equivalent of one tree for each British citizen every year. Eighty percent of all aluminium could be recycled. When you consider that manufacturing one aluminium soft-drink can represents a use of energy equivalent to burning half a can of gasoline, the level of waste and inefficiency is obvious.

Waste and inefficiency are major reasons why we are faced with global warming.

Indeed, as Ted Trainer points out, it is the over-consumption of the North that consigns the people of the Third World to just this fate. Three-quarters of the world's resources are used by the industrialized world, a quarter of the world's population. And many of these resources are acquired on the cheap, mined and harvested by workers who are paid a small fraction of corresponding rates in the North.

The people of the Third World subsidize the consumption patterns of the affluent North.

The gap between the Third World and industrialized nations
Source: US Presidential Commission on World Hunger/The Hunger Project

Consider *three meat meals a day*

The North overconsumes. Ted Trainer, of the University of New South Wales, writes in his book *Developed to Death*: 'Unless you are prepared to make wild assumptions about resources and technology, the inescapable conclusion is that there is no chance of all people in the world living as affluently as people in the rich countries do at present. We are the overdeveloped countries and the rest are the never-to-be-developed countries.' The one billion people living in the industrialized world consume twice as much meat as the three billion people living in the Third World. A US citizen eats, on average, over 100 times as much meat as the typical citizen of India. It is impossible to justify consumption levels in the industrialized world way above those needed to provide life's basic essentials whilst people elsewhere die of starvation and neglect.

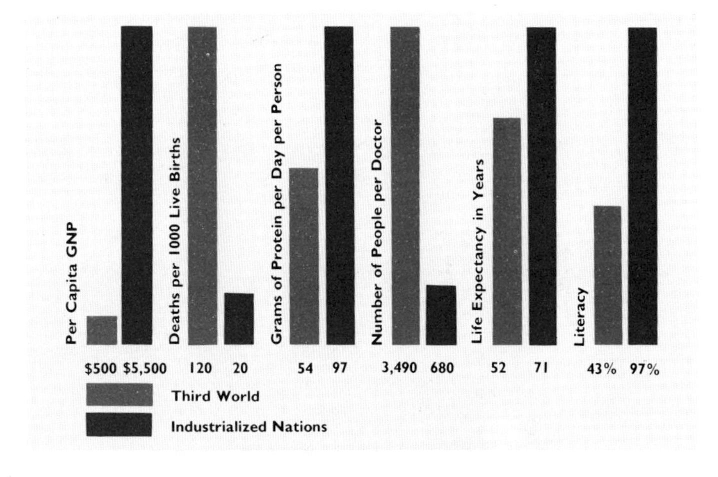

'The great hope for civilization lies in the fact that people can recognize how the human predicament evolved and what changes need to be made to resolve it. No miracles, no outside intervention and no new inventions are required. Human beings already have the power to preserve the Earth that everyone wants – they simply have to be willing to exercise it.'

Anne and Paul Ehrlich, *Earth*

***Consider** the costs of maintaining the world's military establishments*

That the world spends over $2.5 billion *a day* on its military establishments, dwarfing into insignificance expenditure on environmental protection, health care and development aid, is contemptible. The 'Tropical Forest Action Plan', intended to preserve this invaluable resource, would cost $1.3 billion a year over the course of a few years – only *half a day* of military expenditure annually. *Two days* of military spending would finance the 'United Nations Action Plan' to halt Third World desertification for *20 years*. Equipping every Third World household with an energy-efficient cooking stove, limiting global warming and relieving the fuelwood crisis, would cost $1 billion – the same amount the US Government paid the Philippines in 1988 for continued use of their military bases at Subic Bay and Clarke Base.

But the cost of militarization is measured not just in economic terms. As Murray Weidenbaum, chair of President Reagan's Council of Economic Advisors in the early 1980s, has observed, it 'can be more meaningfully expressed in thousands of men and women pulled away . . . from civilian pursuits, millions of man-years of industrial effort, millions of barrels of oil pumped from the earth [enhancing

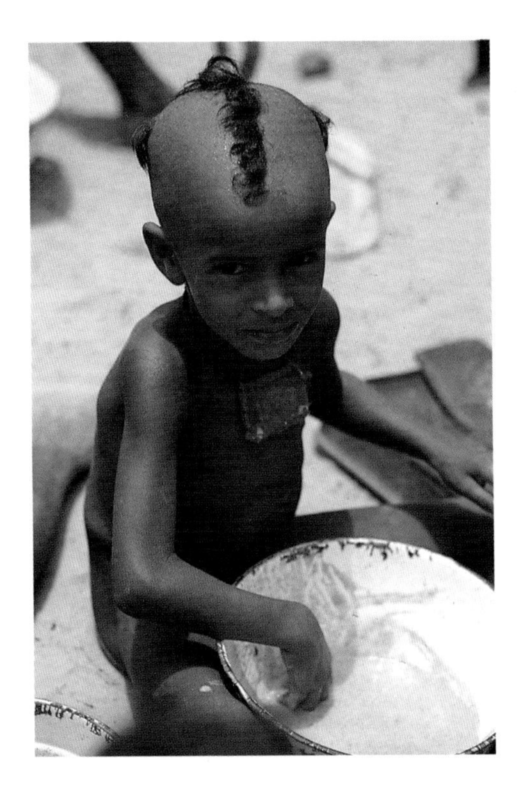

global warming], and thousands of square yards of planet space filled with equipment and debris.' Even in terms of military goals, this diversion of resources makes no sense. What use is a cruise missile if global security is threatened by a disrupted climate? 'Pursuing military security at the cost of social, economic and environmental well-being is akin to dismantling a house to salvage materials in order to erect a fence around it', writes Michael Renner in *State of the World 1989*.

It is misplaced priorities and false perceptions of what constitutes security that give rise to the greenhouse problem, diverting economic, natural, technical and human resources away from a far more pressing defence of our long-term interests.

Finally, consider . . .

The devastation of the tropical forests, one of the world's most diverse and bountiful resources. Susan George, author of *A Fate Worse than Debt*, considers that 'the contributions of huge foreign loans to environmental plunder, widespread impoverishment and ethnocide can no longer be denied.'

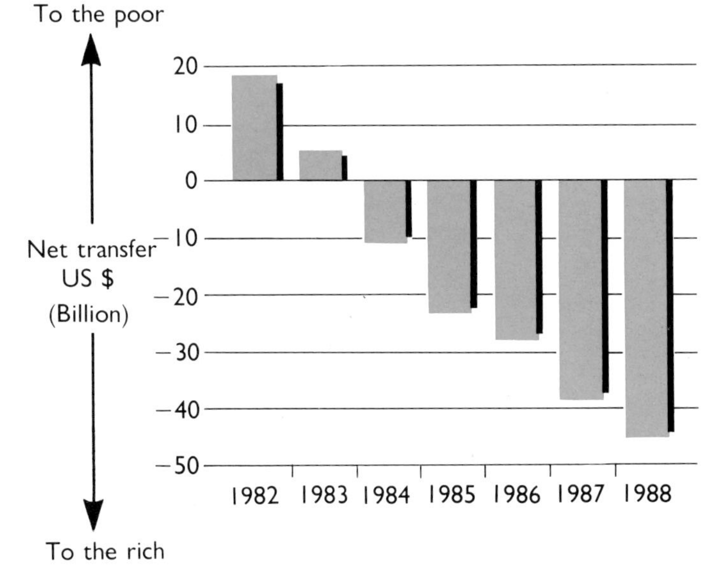

*The flow of money between the North and the Third World. The net transfer to the Third World is the income from new loans minus interest payments on old loans and repayment of capital. Source: World Bank/*The Guardian

'Small farmers are held responsible for environmental destruction as if they had a choice of resources to depend on for their livelihood, when they really don't. In the context of basic survival, today's needs tend to overshadow consideration for the environmental future. It is poverty that is responsible for the destruction of natural resources, not the poor.'

Geoffrey Bruce,
Canadian International Development Agency

'For the head of a poor family like myself and many others, there's no alternative [but to burn the forest]. We have no land to grow food and those who do have land are often not allowed to use it by the landlord, not even to grow potatoes or maize for the family. Therefore we live this life of slavery, burning charcoal, working for others, always remaining poor.'

Charcoal burner, interviewed in
'Death in the Rainforest' (BBC TV)

So much environmental destruction is forced by the demands placed on Third World governments to increase exports in order to meet debt repayments and compensate for falling commodity prices. As we have already noted, clearance rarely provides food for local markets – production is primarily for export. Cattle ranching, for example, has resulted in the loss of over two million hectares a year of the Central American and Amazonian rainforest, catering to the western demand for cheap beef. The poor do not benefit from this process, but they are forced to destroy the forest to fund great national debts.

Something has gone terribly wrong with our world food system.

Today, more than enough food is produced to feed the entire population of this planet — yet millions of women, men and children are forced to go hungry every day. AND their numbers are rising. Food — the most basic human need — is increasingly beyond the reach of the poor and hungry. There's no shortage of food in the world today — but there is a shortage of justice!

What is our role in this crisis? And what can we do about it?

Oxfam has identified 5 main areas in which we in the UK are involved in the crisis of world hunger.

D — DEBT

To the poor in the Third World, the international debt crisis is not something to read about in the newspapers but a matter of survival.

With very high interest rates in the rich countries — poor countries struggling under huge foreign debts cannot afford to pay the interest let alone repay the loans. Third World countries are forced to take drastic austerity measures in order to receive IMF "help" and it's the poor who lose out. More and more keenly needed resources are diverted from things like health services to go to paying off the debt. Brazil is saddled with the Third World's largest debt at over $90 billion — 2½ times as much as its government is spending on its annual budget — on social services, on agriculture, on education, on industry — on everything! The pattern is repeated throughout the Third World. The real victims of the debt crisis are the poor.

WE WANT THE BURDEN OF THE DEBT CRISIS AND CRIPPLING INTEREST RATES REMOVED FROM THE SHOULDERS OF THE POOR. THE POOR SHOULD NO LONGER HAVE TO PAY THE PRICE FOR A CRISIS THEY DID NOT CREATE.

A — AID

Development aid is declining. The British government's aid budget has been cut back while many other Western governments have increased aid. In spite of a slight recent increase, our aid budget is worth much less in real terms than the 1979 aid programme. Each year Britain spends more on buying, storing and disposing of agricultural surpluses than it does on overseas aid, yet the need for aid has never been greater.

The quality of aid is just as important as the quantity. Aid should reflect the needs of the poor and not be dominated by self interest. Much aid goes on political grounds to political allies — the Falklands get more than the major countries of Africa's Sahel region — and to obtaining export orders for British companies.

WE WANT AID TO BE INCREASED AND REFORMED SO THAT IT REALLY REACHES THE POOR.

T — TRADE

More and more land in an increasing number of Third World countries has been turned over to grow crops like coffee and peanuts for export to Europe and the USA. Four countries in Africa are dependent on just one crop for over 70% of their income. Take Chad — four-fifths of its export earnings come from cotton. In an attempt to pay off large debts these compete with each other in the world market place. And all the time commodity prices are falling — according to a World Bank report African commodity prices dropped 27% from 1980-83. Yet prices of the imports many poor countries need go on spiralling upwards — by 1981 it took one Latin America country about ten times as much beef to buy one barrel of oil as it did in 1973. So even if Third World countries do produce more — this still results in them getting less!

WE WANT THE TERMS OF TRADE IMPROVED SO THAT THE POOR ARE NO LONGER STRANGLED BY THE WORLD MARKETPLACE, SO THAT THEY GET FAIR PRICES FOR THE PRODUCE ON WHICH THEY DEPEND, AND SO THEY CAN SELL A WIDER RANGE OF GOODS IN OUR MARKETS.

A — AGRICULTURAL POLICIES

Agricultural policies pursued by the wealthier countries add to the scandal of hunger. The EEC has encouraged the production of massive food surpluses — food mountains. Some of these we send as food aid to hungry people who in turn are exporting nutritious food they grow not to feed themselves but to feed our livestock — for example soya from Brazil and anchovies from Peru.

Take the island of Negros in the Philippines — known as "Sugarlandia" Negros produces 70% of sugar in the Philippines; invaluable foreign exchange earnings for a country suffering a huge external debt. But Negros, however, now has a new name — the Ethiopia of South-East Asia. A collapse in the world price of sugar in 1984/85 contributed to thousands of sugar workers being laid off. An estimated thirty per cent of the island's children are suffering from malnutrition as a result.

How did this come about? The EEC has over stimulated the production of sugar to the point where each year it produces 3 million tonnes more than it consumes. This is dumped on the world market and has been a major cause of the slump in world prices.

WE WANT EEC POLICIES CHANGED SO THAT THE WORLD'S POOR ARE NOT HAVING TO SUPPORT OUR WASTEFUL COMMON AGRICULTURAL POLICY. WE WANT THERE TO BE SPACE IN WORLD AGRICULTURAL COMMODITY MARKETS FOR THOSE WHO NEED IT MOST.

A — ARMS

The poor are becoming poorer, hungrier and increasingly angrier and more desperate. What was once a struggle for justice and a fairer distribution of food, land and power is becoming a chaotic fight for survival.

So far in the 1980s there have been food riots in the Dominican Republic and Haiti, gangs of Brazilian mothers raiding supermarkets in the search for food for their children and street demonstrations demanding food in Mexico.

Governments are responding by importing arms — over £28.2 billion worth each year — many intended for use against their own hungry people. Three-quarters of Britain's arms exports go to the Third World including large orders for crowd and riot control equipment.

WE WANT A TRANSFER OF SPENDING FROM THE ARMS RACE TO DEVELOPMENT — SECURITY THROUGH A HUNGER-FREE WORLD NOT SECURITY THROUGH MORE AND MORE SELLING OF ARMS.

1985 FIGURES

CORNWALL COLLEGE LIBRARY

How did such debts arise? The banks of the North faced a problem during the 1970s. Accumulating massive funds as oil prices rose, they needed to get this capital out into the world, earning interest. As Christian Aid observed in their publication *Banking on the Poor*. 'Self-interest is a driving force in the Third World debt crisis. The initial lending by the private banks in the 1970s was motivated not by altruism but by self-interest. The money lent by the banks had in turn been generated by the self-interest of the oil-exporting countries brought together in OPEC.' The Third World represented an untapped market and the banks enthusiastically lent money, often with scant regard for the ability of their debtors to repay. Many could not. The slump in global economic activity that followed, the depression in commodity prices and Third World export earnings and the high interest rates of the early 1980s severely limited the ability of even the richest nations of the Third World to meet their repayment schedules.

The original loans were intended to finance development, as capital investment for industry, agriculture and other sectors, which would lead to economic growth and the ability to repay. In reality, the money was largely wasted – on overambitious schemes for industrialization based on western models which ignored local conditions; on graft and corruption; on guns rather than butter. And new loans were taken out to pay off the old. The poor, the rural population, rarely benefited. It was considered that financing large-scale industrial development would ultimately result in the 'trickle-down' of benefits as overall economic health improved. As development plans failed, the poor were left without any reward. Now they bear the brunt of the austerity programmes forced on Third World governments by organizations such as the International Monetary Fund (IMF) to ensure some degree of debt repayment.

The economic policies of the North add directly to the debt bill. The increasing of interest rates to combat inflation has been a common policy in many industrialized nations. During the mid-1980s, every one percent rise in US interest rates increased Third World debt by $4 billion. While such policies may assist northern economies, they condemn the Third World to greater suffering.

At present, over $40 billion a year is transferred from the poor of the Third World to the North as a result of the misguided policies which produced and prolong the debt crisis. It is no wonder that Third World nations plunder their resources, condemning the North as well as themselves to environmental disaster. It is the inequality built into the world's socio-economic system that is causing the greenhouse problem. When one considers the many ways in which socio-economic forces deprive the Third World and favour the North, it is difficult to avoid the conclusion that some kind of conspiracy is going on.

Again and again, the rich are getting richer and the poor are getting poorer. Whether this travesty of justice occurs by accident or design, there is no doubt that the path of development we are following is heading inexorably towards a tragedy of global dimensions – and that we must correct this course or face a threatening future.

OUR COMMON FUTURE

As Gro Harlem Brundtland has observed: 'We in the North have a special responsibility . . . For too long we have exported our first generation of environmental problems to the Third World and maintained an economic system which leads to environmental decline in developing countries. . . . It is time that we realize that we all share a common future.'

In 1983, the Secretary-General of the United Nations (UN) called on Gro Harlem Brundtland, the Prime Minister of Norway, to chair a special committee to formulate a 'global agenda for change'. The main aim of the committee – the World Commission on Environment and Development (WCED) – was to propose long-term environmental strategies for achieving sustainable development by the year 2000 and beyond. Sustainable development has been defined in a variety of ways. According to the WCED, it is development which meets the needs of the present without compromising the ability of future generations to meet their own needs. If we deplete a resource

'Sustainable development is development that meets the needs of the present without compromising the ability of future generations to meet their own needs. It contains within it two key concepts:

- the concept of 'needs', in particular the essential needs of the world's poor, to which overriding priority should be given, and
- the idea of limitations imposed by the state of technology and social organization on the environment's ability to meet present and future needs.

[T]he notion of physical sustainability implies a concern for social equity between generations, a concern that logically must be extended to equity within each generation.'

World Commission on Environment and Development, *Our Common Future*

such as the rainforest, where will our children and their children find the medicines they need? If we change global climate by polluting the atmosphere, will the environment be able to support the growing population of the planet? Sustainable development is development which takes a broader view than immediate profit; balancing short-term interests against long-term consequences.

The WCED reported in 1987. *Our Common Future* documents, in dry, technical terms, the many ways in which society is failing to meet the needs of millions of people throughout the world in the present-day and is prejudicing the survival chances of future generations.

'Many critical survival issues are related to uneven development, poverty and population growth. They all place unprecedented pressure on the planet's lands, waters, forests, and other natural resources, not least in the developing countries. The downward spiral of poverty and environmental degradation is a waste of opportunities and of resources. In particular, it is a waste of human resources.... What is needed now is a new era of economic growth – growth that is forceful and at the same time socially and environmentally sustainable.'

We will look at how this might be achieved in later sections. For the time being, we shall consider what might happen if we fail in the quest for sustainable development.

FUTURE SHOCK

By the early decades of the twenty-first century, within the lifetime of most of us, the world will have changed once more. If projections based on the extrapolation of current rates of change are correct, there will be more than 8 billion people living on the planet by the year 2030. Agriculture will have expanded to meet this increase in population. The world's forests may have been near-completely destroyed. And energy demand will have more than doubled.

We are already ill-equipped to meet the demands of the Earth's current population. According to *Ending Hunger* (Praeger, 1985), compiled by the Hunger Project: 'More than 1 billion people are chronically hungry. Every year 13 to 18 million people die as a result of hunger and starvation. Every 24 hours, 35,000 human beings die as a result of hunger and starvation – 24 every minute, 18 of whom are children under five years of age . . . The worst earthquake in modern history – in China in 1976 – killed 242,000 people. Hunger kills that many people every seven days.'

As population rises and climate changes, the situation can only get worse.

By the year 2030, if no action is taken to slow the buildup of greenhouse gases, then their combined heat-trapping effect will be the same as if we had doubled the concentration of carbon dioxide from its natural level of 280 parts per million. According to the latest estimates, that will generate warming of up to 3°C – and more warming will follow as the oceans slowly respond to the change in atmospheric composition, even if the buildup of greenhouse gases stops then. At the end of the most recent ice age, the average temperature of our planet rose by about 4°C, over a span of several thousand years. Greenhouse gases released by human activities seem set to warm the globe by roughly the same amount, but over a span of only a few decades – less than a human lifetime. Sea level will rise by up to 1m as the oceans expand and glaciers melt, 20 times the rate of change experienced this century. And temperatures will continue to rise with no respite unless the release of greenhouse gases is brought under control. The heat trap will continue to increase in intensity, taking global temperatures up by 8°C or more by the end of the twenty-first century. The last time the world was that warm it was dominated by the dinosaurs, not by humankind.

We now look in more detail at how climate might change as global warming progresses, and at what these changes will mean for humanity, and for the entire living world.

2

THE GLOBAL GREENHOUSE

People often think that the greenhouse effect is something controversial, a scientific theory that is not proven and still the subject of debate. But this is not the case. Functioning naturally, the greenhouse effect explains why our planet is the way it is, and why it provides a suitable home for life. Greenhouse gases trap heat near the surface of the globe that would otherwise escape into space. They make the world warmer than it would be if there were no blanket of gases – no atmosphere – surrounding it. Natural greenhouse gases in the atmosphere, especially carbon dioxide and water vapour, make our planet some 35°C warmer than it would be if there were no greenhouse effect at work. Without these gases, the Earth would be a frozen, lifeless ball in space.

Stephen Schneider, of the National Center for Atmospheric Research, points out that the greenhouse effect accounts for a lot of the phenomena we observe in our own Solar System. When astronomers look up to the heavens, they see a red planet, Mars, which has a very thin atmosphere with a weak greenhouse effect, and very low temperatures like those in Antarctica. They also see a bright, white planet, Venus, which has a very thick atmosphere with a hundred times the amount of gas in the Earth's atmosphere, and a very strong greenhouse effect that makes it hotter than a baker's oven. But,

The planet Mars

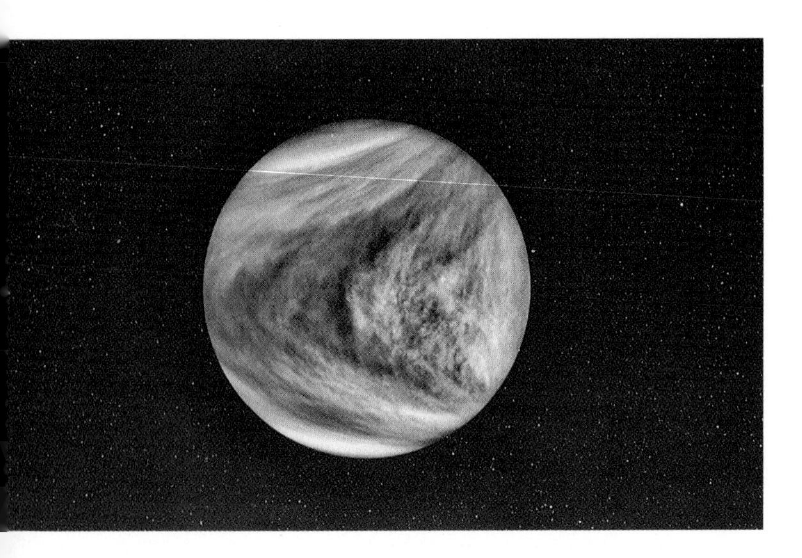

The planet Venus

says Schneider, the Earth is 'just right – it's the planetary Goldilocks phenomenon, and it's explained by the greenhouse effect.'

It works like this. Light is a form of energy, and carries heat. This electromagnetic energy covers a band of wavelengths, known as the spectrum. Visible light covers the range from red, at one end of the visible part of the spectrum, to violet, at the other end. But there is also radiation at longer wavelengths (beyond the red) and at shorter wavelengths (beyond the violet), which our eyes cannot see. Energy from the Sun in the form of electromagnetic radiation – visible light and some ultraviolet – passes down through the atmosphere of the Earth almost unhindered. Some is reflected back into space and lost. A small proportion, chiefly the ultraviolet, is absorbed by gases in the atmosphere. But, apart from the interception of the ultraviolet part of the spectrum by the ozone layer high in the atmosphere, most of the energy passes straight through.

The energy that reaches the surface of the Earth is almost all in the form of visible light, at relatively short wavelengths. On striking the Earth, it heats the surface of the ground and the sea. The warm surface itself radiates energy upwards through the atmosphere and back out into space. But this departing energy is different from the incoming sunlight in one crucial respect – it has a longer wavelength, beyond the red end of the visible spectrum.

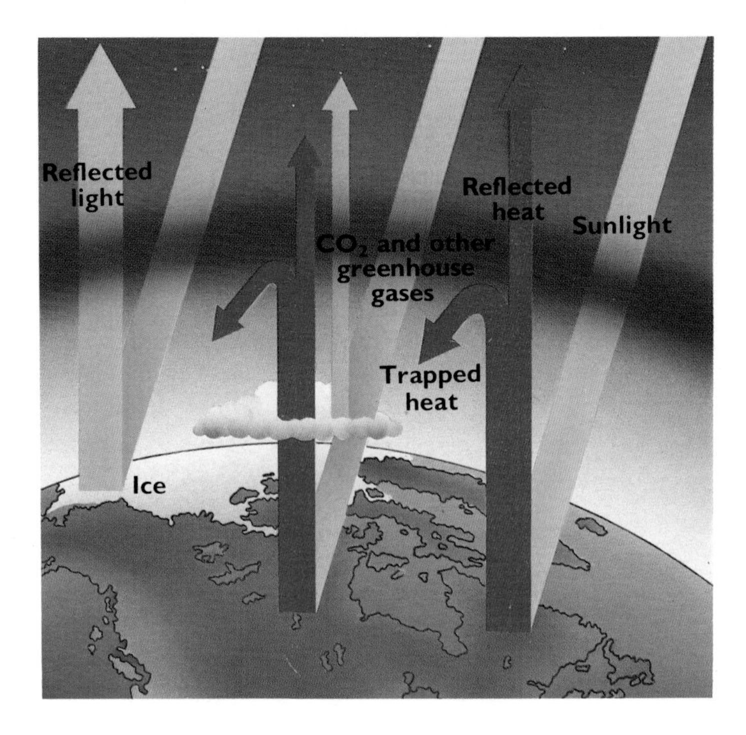

The greenhouse effect
Source: Fortune

The wavelength of the radiation emitted by any object depends on its temperature. A lump of iron appearing 'white' hot, for example, is hotter than a lump appearing 'red' hot. It radiates light with short wavelengths. The surface of the Sun is very hot, about 6000°C, so it radiates very short wavelength radiation, visible light. The surface of the Earth is much cooler, on average about 15°C, so it radiates energy at much longer wavelengths, mostly in the infrared part of the spectrum outside the range of visible light. An object with a temperature of 15°C feels warm because it emits radiation, but we cannot see the radiation that it emits.

Although gases like carbon dioxide and water vapour are transparent to visible light – like the panes of a greenhouse – and let energy from the Sun in, they do not allow infrared radiation to pass through. This is the heat trap: some of the heat radiated by the surface of the Earth is trapped near the ground and keeps the surface of our planet warmer than it would otherwise be.

So the basic mechanism of the natural greenhouse effect is well understood. It is not controversial at all. Nor is there any doubt that increasing the concentration of greenhouse gases in the atmosphere will make the world warmer. There is some uncertainty, however, as to how climate will vary as the composition of the atmosphere changes. How fast will temperature rise? Are there processes which will counteract the warming? How will rainfall levels vary? But even so, the broad picture – the nature of the threat – is clear.

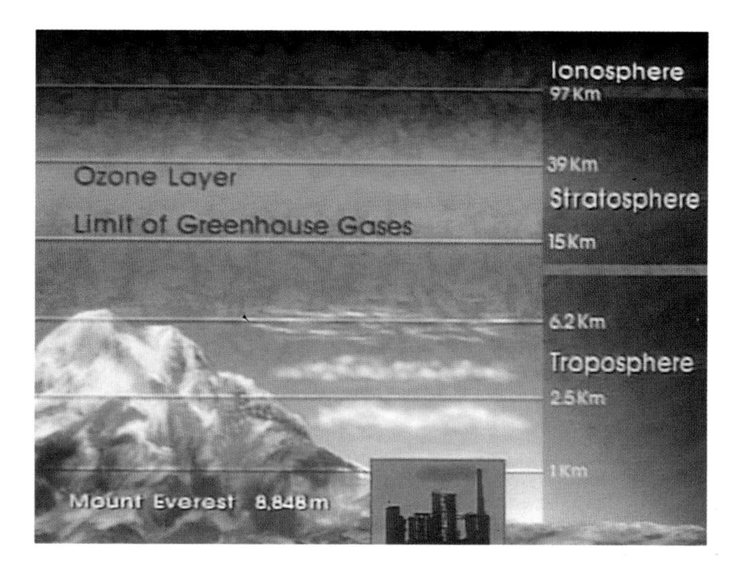

Layers of the Earth's atmosphere

The greenhouse effect has been the subject of two major reviews by the scientific community in recent years. Drawing on the skills of the world's climate experts, these were undertaken by the US Department of Energy and by SCOPE. SCOPE stands for the Scientific Committee on Problems of the Environment. It is a standing committee of the International Council of Scientific Unions, a coordinating body for the major scientific organizations of the world. SCOPE is, in effect, the environmental watchdog of the scientific community. Reviewing the evidence of the threat posed by global warming and taking full account of the uncertainties still to be resolved, the SCOPE editors concluded that the greenhouse effect should be considered 'one of today's most important long-term environmental problems.'

MODELLING A WARMER WORLD

Predictions of the way in which the climate will respond to the greenhouse effect are made using climate models running on powerful electronic computers. These are not models in the everyday sense of the word, constructions of wood or plastic and glue which represent in miniature something like an aircraft or a ship. They are sets of mathematical equations that describe the basic laws of physics at work in the atmosphere, and the way in which physical processes interact with one another to shape our climate. Like a model aircraft that you hang from the ceiling or a ship in a bottle, they represent something that is too big to 'play' with directly – in this case, the whole world.

Climatologists cannot take a whole planet and change the composition of the atmosphere to see how that affects climate. But they can calculate the effects, by altering the numbers in the computer model that represents the composition of the atmosphere, the brightness of the Sun or any other influence on climate. In this way, scientists

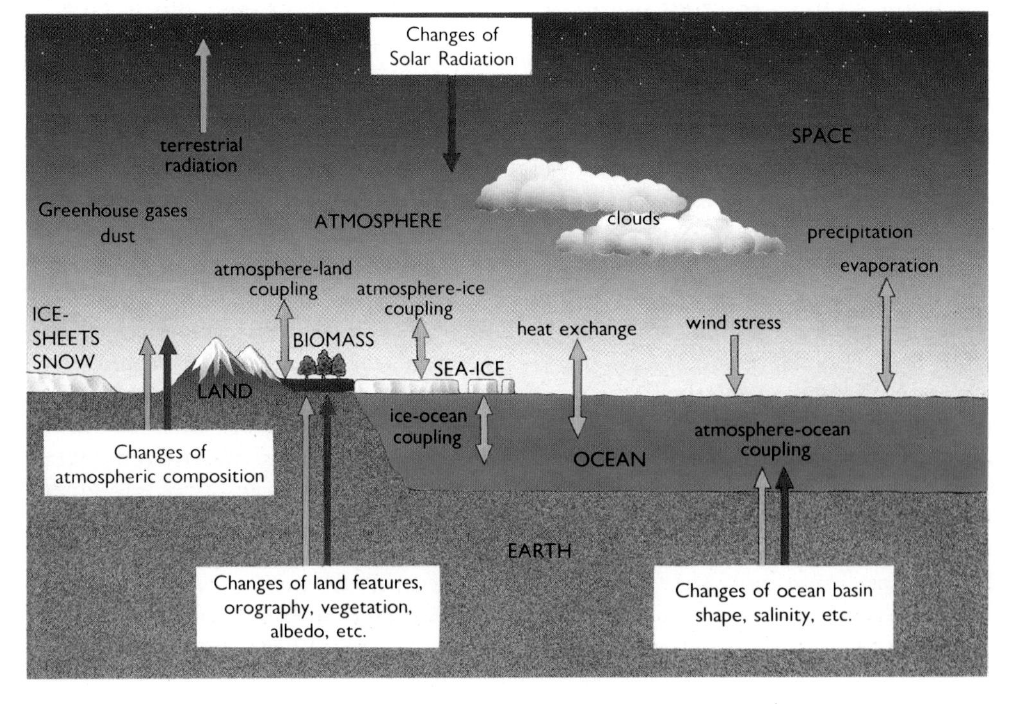

The many different processes of the climate system that have to be modelled in order to predict global warming. The red arrows are examples of external processes which cause climatic change. The yellow arrows are examples of internal processes. Source: Global Atmosphere Research Programme

can get some idea of the outcome of the experiment that we are running in the real world, providing a timely warning of the consequences of our actions – before we see the results by looking out of the window.

The simplest climate models deal only with basic physics, and treat the surface of the Earth as a smooth, uniform globe. These models do not need to run on a powerful computer – they could be used on a games machine at home – and can be run again to test their results. James Lovelock, author of *The Ages of Gaia* (OUP, 1988), has compared this kind of model to the simplest abstract painting, in which a few strokes of the brush capture the essence of the object to be portrayed. Simple, but effective. Even these models are sufficiently accurate to calculate the amount of global warming produced by the natural amount of greenhouse gases in the atmosphere. We know the calculations are correct by comparing the results with actual temperature from Earth, Mars, Venus and our airless Moon. But they do omit crucial details, particularly those needed to predict the regional impact of global warming.

More complex models include more detailed physical processes – interactions between the oceans and the atmosphere, changes in cloud patterns – and more realistic geography – mountains and differences between land and sea. These are known as General Circulation Models (GCMs). They are closely related to the computer models used to predict the weather from day to day. Good GCMs run only on the biggest supercomputers, and need many hours of computer time to calculate how the weather will respond to a change in any of the factors that influence our climate.

Yet there is only so much information that can be included on a finite canvas, or in a present-day computer – far less than the myriad influences on the real world. This is why even the best climate model is not very good at predicting regional changes in climate – the way in which global warming will affect, say, the UK as opposed to the neighbouring Netherlands. It is impossible to include a representation of all the processes which affect climate at this level. The climate modeller paints with a broad brush.

Models are, however, good at explaining gross differences in climate. They can calculate, correctly, the difference in temperature between an airless globe and the real Earth. They can estimate the difference in wind patterns between an ice age and the present day. And they are very good at calculating the differences between the seasons. This is by no means straightforward. The temperature difference between winter and summer in the Northern Hemisphere is about 15°C, but the corresponding variation in the Southern Hemisphere is only about 5°C. We know why this is so – it is because there is more land in the north, and land changes temperature more quickly than ocean. The best GCMs can calculate this

Planet Earth

pattern of change exactly, and this represents a real triumph for the climatologists who have developed these models. They have captured the complexity of the climate system in thousands of lines of computer code.

But the temperature rise likely over coming decades as global warming develops is much less than the change from winter to summer, only a few degrees change, as opposed to ten or more. It is much harder for a GCM to predict changes on this scale. An error of a degree in calculating a change of 15°C is acceptable, less than ten percent. An error of a degree in calculating a change of 1.5°C is over 65 percent. And understanding of the mechanisms which will determine the rate of warming is not as complete as knowledge of the processes which determine the shift from summer to winter.

As temperatures rise, 'feedback' effects will determine the ultimate rate of warming. Feedback works like this. Higher temperatures near to the poles will melt snow and ice, uncovering bare ground which absorbs more heat. This will amplify the initial temperature change. The amount of water in the atmosphere will increase as the oceans warm and more seawater evaporates into the air. Water vapour is a greenhouse gas, so again this will have an amplifying effect. These are positive feedbacks. Something causes a change in climate, and that change feeds back upon itself, amplifying the change. There are also negative feedbacks. They tend to reduce the impact of an initial change. Cloud patterns will change as the world warms. Some types of cloud reinforce the greenhouse effect, others reduce it. And there are many more feedback mechanisms. There is quite a lot of uncertainty about these crucial feedback processes. Warming will occur – but will it be at a rate of 0.10°C every ten years or 1°C every ten years? This uncertainty is the reason why predictions of the rate of global warming are generally given as a range. No-one seriously argues that feedback will totally counteract the warming, but it could reduce it by a significant fraction or – more likely – amplify it over and above the initial change.

So what do we know for certain?

In order to look to the future, we have to look back at the climate of recent decades for evidence that the existing buildup of greenhouse gases has affected the climate. The record of global air temperature is a good starting point. It is the most commonly used measure of the state of the climate sys-

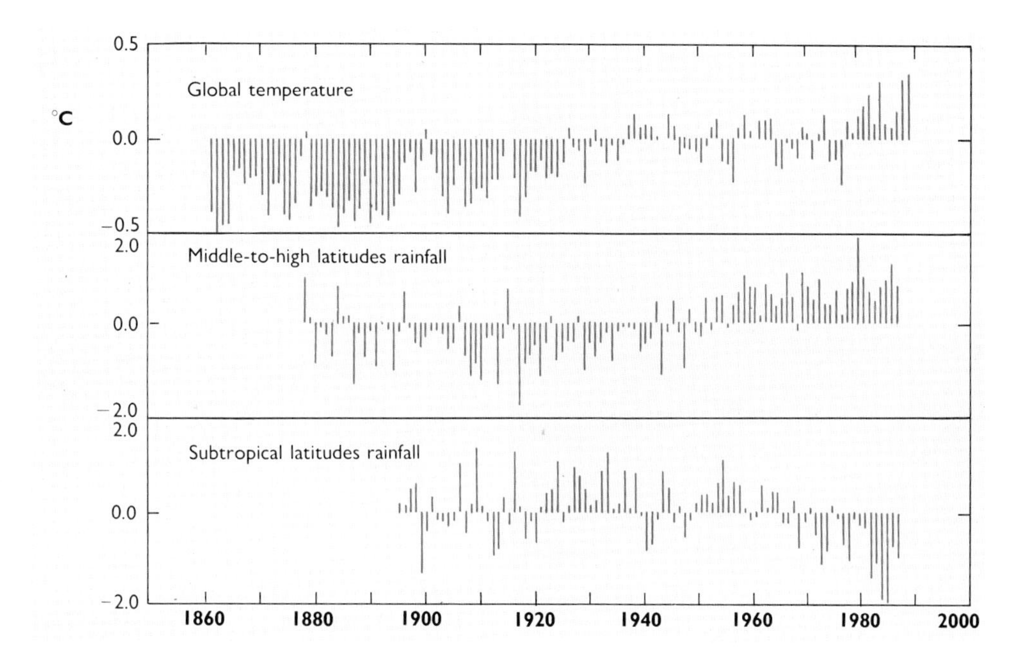

Recent trends in climate. The top graph shows surface air temperature averaged worldwide for each year since 1861. The data are expressed as departures from the 1951–79 mean. The lower two graphs show rainfall amounts averaged over the land areas of middle-to-high and subtropical latitudes of the Northern hemisphere. Because rainfall amounts vary widely from place to place and from season to season, a simple index of rainfall amount which takes these factors into account has been used. The units are arbitrary

tem and the variable responding most directly to the greenhouse effect.

Climatologists estimate the temperature of the planet from the many observations taken around the world each day, a task coordinated by the World Meteorological Organization. Based on thousands of recordings taken at land stations, supplemented by data taken by ships at sea, the record of global temperature has improved substantially over recent years. And as a result of painstaking detective work, tracking down early observations and eliminating errors due to instrument changes, the surprising amount of data available for the nineteenth century is now being interpreted. Much of it was lost in archives and libraries for many years but is now made available for global warming studies.

The computer models of the climate system tell us that the strengthening of the greenhouse effect that has occurred over the past 100 years should have resulted in global warming of between 0.3 and 1.1°C. The world has, in fact,

warmed by about 0.5°C during this period – well within the range of the model predictions.

This is strong support for such predictions, but does not prove beyond doubt that the greenhouse effect is responsible for the observed warming. Climate can change for many reasons. Large volcanic eruptions throw enough pollution into the atmosphere to block out part of the Sun's heat, cooling the world by a fraction of a degree for a year or two. The energy output by the Sun may vary, producing a change in global temperature. The temperature of the oceans may alter as current patterns change, affecting the heating of the overlying air and generating climatic change.

Is it possible that some other factor, or combination of factors, has given rise to the twentieth century warming, not the greenhouse effect? The answer must be: yes, it is *possible* – but it is *unlikely*. An unparalleled rise in temperature is occurring at the same time as we are polluting the atmosphere with greenhouse gases at an ever-increasing rate. To argue that it is all coincidence really is whistling in the dark. The most plausible explanation for the changes observed over the past 100 years is that they are largely due to the greenhouse effect, with other natural causes of climatic change obscuring – offsetting or amplifying – the underlying warming from time to time. This is why the rate of warming has varied throughout the past century. The temperature of the planet rose quickly during the 1920s and 1930s but then marked time during the 1940s and 1950s, staying roughly constant. Since 1970, warming has set in once again with renewed vigour.

The underlying warming will only stand out clearly when the change is going faster that anything caused by natural mechanisms – when the 'signal' rises above the 'noise', to use the jargon. We are not quite at this point yet, which is why scientists tend to be cautious when asked whether or not the greenhouse effect is already upon us. For similar reasons, they cannot say for sure whether or not individual extreme events such as the North American drought of 1988 or the mild winter that followed in western Europe are attributable to the greenhouse effect, except with the benefit of many years' hindsight.

Scientific uncertainty is often used as a reason, or as an excuse, for delaying action to deal with environmental problems. As Richard Benedick, US Deputy Assistant Secretary for the Environment, has observed, 'the notion that "future risk cannot be estimated with certainty" is sometimes translated, by a curious leap of faith, into the assertion that "there is no future risk"'. In March 1989, Mostafa Tolba, head of the United Nations Environment Programme, responded to reservations concerning the scientific case on ozone depletion, with the sharp comment: 'I must emphasize that we cannot allow doubt to delay action. And I must state that the best available scientific information indicates that by the 1990s the consequences will begin to be felt everywhere.'

If scientists cannot be absolutely sure till after the event, perhaps we should assume the worst, assume guilt till innocence is proven, rather than otherwise. We are not in a court of law where the interests of the accused must be put foremost, innocence assumed and proof required to demonstrate certainty of

guilt. It is the interests of humanity, rather than of scientific proof, that must be placed first. Given the likely consequences of the greenhouse effect, we must embark on the safest course of action. And that means we should act now.

The possible consequences of further global warming are clearly demonstrated in the results of recent analyses of large-scale change in rainfall patterns. During the mid-1980s, a team of scientists from the University of East Anglia in the UK and the University of Massachusetts and NOAA, the National Oceanic and Atmospheric Administration, in the US undertook the first, comprehensive review of trends in rainfall around the world. One of us was a member of that team, and the results of the study led to many sleepless nights.

Computer models predict that when the greenhouse effect takes hold there will be more rain and snowfall overall at higher altitudes. There are also indications that the dry areas near the equator will become drier. This is precisely the pattern the team saw occurring in the real world over recent decades. A pronounced trend towards more rain and snowfall was evident at higher latitudes of the Northern Hemisphere since the 1950s and a downward trend in rainfall – the mirror image – in subtropical latitudes. The discovery came as a shock. Everyone knew that drought had been affecting the Sahel, the long band of semi-arid countries just south of the Sahara desert, and Ethiopia to the east, since the late 1960s; but no-one realized that rainfall *throughout* the subtropics had been declining steadily for four decades, since the time global temperature peaked after the first major warming episode of the twentieth century.

Pictures of the tragic famine in Ethiopia brought home just what the computer models were saying about the

impact of global warming on the real world, on the fate of real people. Scientists tend to consider the greenhouse problem in terms of dry statistics, degrees Celsius and millimetres of rainfall, but the impact should really be considered in terms of human life.

What do we know about the ways in which the climate will change over coming decades – and what will this mean for the eight billion people populating the planet in the year 2030?

THE TWENTIETH CENTURY WARMING

The computer models used to predict global warming churn out thousands of statistics – in tables, graphs and charts – detailing the possible state of our future climate. A refinement to a model here, a change in a value there, and more figures flood the laboratory. Yet, surprisingly, there are many details that the different models and different experiments agree on. And, as the consensus emerges, we are gradually developing a portrait of world climate in years to come that is sufficiently detailed to be the basis for a projection of its impact on humanity.

Two kinds of model experiments have been undertaken to determine how climate might change as greenhouse gas concentrations build up. The first set are known as 'equilibrium' experiments. These compare the present-day situation with what might happen when greenhouse gas concentrations reach a certain level, usually the equivalent of twice as much carbon dioxide in the atmosphere as in pre-industrial times. These experiments provide a detailed snapshot of what the world might be like at one point in the future. We will look at these results later. Other 'transient response' experiments give a picture of the way the change in climate evolves with time. They predict the year by year rise in temperature as the atmospheric composition slowly alters.

Since transient model simulations are forecasting the timing of changes in climate, they must take into account the slow response of the oceans to the greenhouse effect. Oceans take more time than the atmosphere to warm – their 'thermal inertia' is high. This is why the sea is often still cold at the height of summer, reaching its highest temperature as the air cools down in early autumn. The thermal inertia of the oceans delays the response of the temperature of the air to any change in atmospheric composition. This lag effect also means that, even if stringent control measures were to bring the buildup of greenhouse gases to a halt tomorrow, temperatures would continue to rise, perhaps for the next 20 years or more, as the oceans slowly respond to today's greenhouse effect. This is another strong justification for immediate action.

'The nineties needs to be a 'turnaround decade'. The looming threats we now face – including climate change, ozone depletion and population growth – have so much

momentum that unless action begins now to reverse them, they will inevitably lead to paralyzingly costly economic consequences and the collapse of social and political institutions.'
Lester Brown, Christopher Flavin and Sandra Postel,
State of the World 1989

The figure (below) shows the results of one transient response experiment, one estimate of how global temperatures might rise as the atmospheric composition changes over coming decades. Developed by Tom Wigley at the University of East Anglia's Climatic Research Unit, it is based on the assumption that current trends in energy consumption, agriculture and so on will continue with no effort made to curb emissions of the greenhouse gases. The warming experienced to date pales into insignificance compared to what lies in store if no counter-measures are taken. The rate of warming accelerates during the 1990s, largely as a result of the marked increase in greenhouse gas emissions that has taken place since the 1950s. Because of the delaying effect of the oceans, we are only just seeing the first impact of that rapid escalation in pollution emissions.

According to this projection, global temperature will lie some 2°C above the present level by the year 2030. When the world warmed by about 4°C at the end of the latest ice age, temperatures reached a peak less than one degree higher than they are today. Geological evidence shows that this warmth lasted for only a few thousand years at most, beginning about 10,000 years ago. Since then, although global temperature has varied slightly as the millennia have passed, there is no evidence that there has ever been a period of sustained

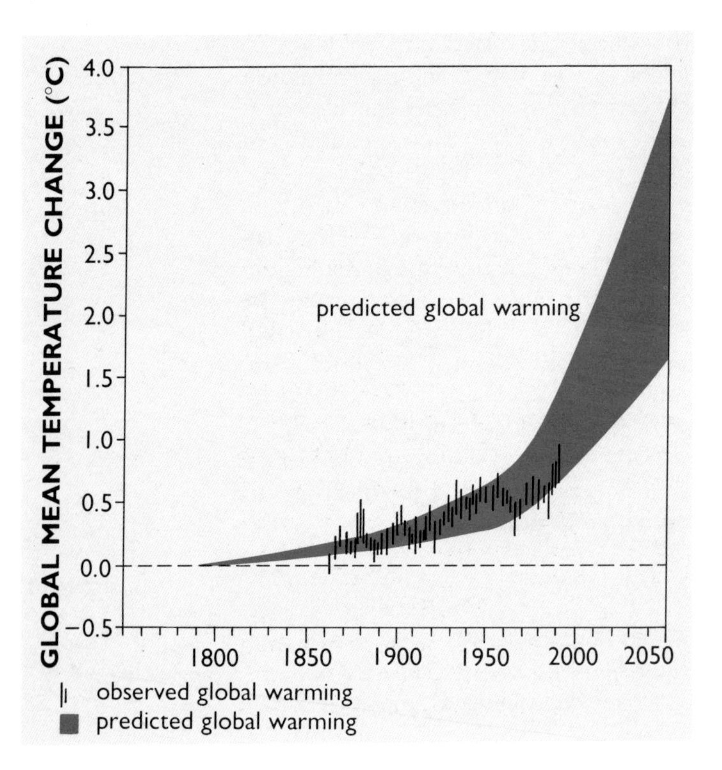

One estimate of the response of global temperature to the strengthening greenhouse effect. The solid line indicates the temperature history of the planet over the past 100 years or so. The shaded area indicates the range within which this experiment suggests temperatures should lie as a result of the increasing amount of greenhouse gases in the atmosphere. The horizontal dashed line shows the temperature level before the Industrial Revolution.
Source: T M L Wigley

warmth hotter than the world is now. By the early twenty-first century, it is likely that the world will be warmer than it has been at any time since the beginning of the most recent ice age, some 100,000 years ago.

Jim Hansèn and colleagues at NASA in the US have undertaken the most sophisticated transient response experiment to date using one of the most advanced GCMs that is available. Published in the prestigious *Journal of Geophysical Research,* it shows how the pattern of temperature change might evolve over the next 60 years. The results for three decades are shown in the figure on the next page. The right-hand column shows the difference between the predicted temperature for these decades and the temperature that would have prevailed had greenhouse gas levels stayed at the level prevailing in 1958. During the 1980s, warming is most evident at high latitudes, near the polar regions, although the Siberian landmass is already warming quite strongly. By the 1990s, warming has extended towards the equator, particularly in the Northern Hemisphere. It is now affecting much of Europe and North America. Two decades on, during the 2010s, only isolated ocean areas remain unaffected and temperatures have risen by up to 2°C over much of the world's landmasses.

The left-hand column of charts shows an estimate of the significance of these trends, calculated by comparing the temperature change with the natural variability in climate – the signal with the noise. A 1°C rise in temperature may be trivial in an area where temperature changes by 5°C from year to year – for example, in the Arctic – but it is very significant in an area where temperatures rarely change by more than 0.5°C – such as the tropics. A value greater than one on this scale can be considered a significant change. Again, the 1990s turns out to be the critical decade. Over much of the world, the rise in temperature could surpass the natural variability in climate sometime within the next ten years.

With global temperature now at a record high and the rate of warming set to accelerate markedly over the next ten years or so, we are undeniably at a threshold.

THROUGH A MURKY CRYSTAL BALL

Even the best computer models cannot forecast exactly what the climate of the world will be like in the twenty-first century. Apart from uncertainties in the models themselves, any prediction must take into account the rate at which the world's population will grow and related trends in agriculture, industrial activity, energy consumption, and so on. And these trends may well be affected by growing popular concern about global warming.

They will depend on unanticipated developments – new inventions and unforeseen political crises. Attention was

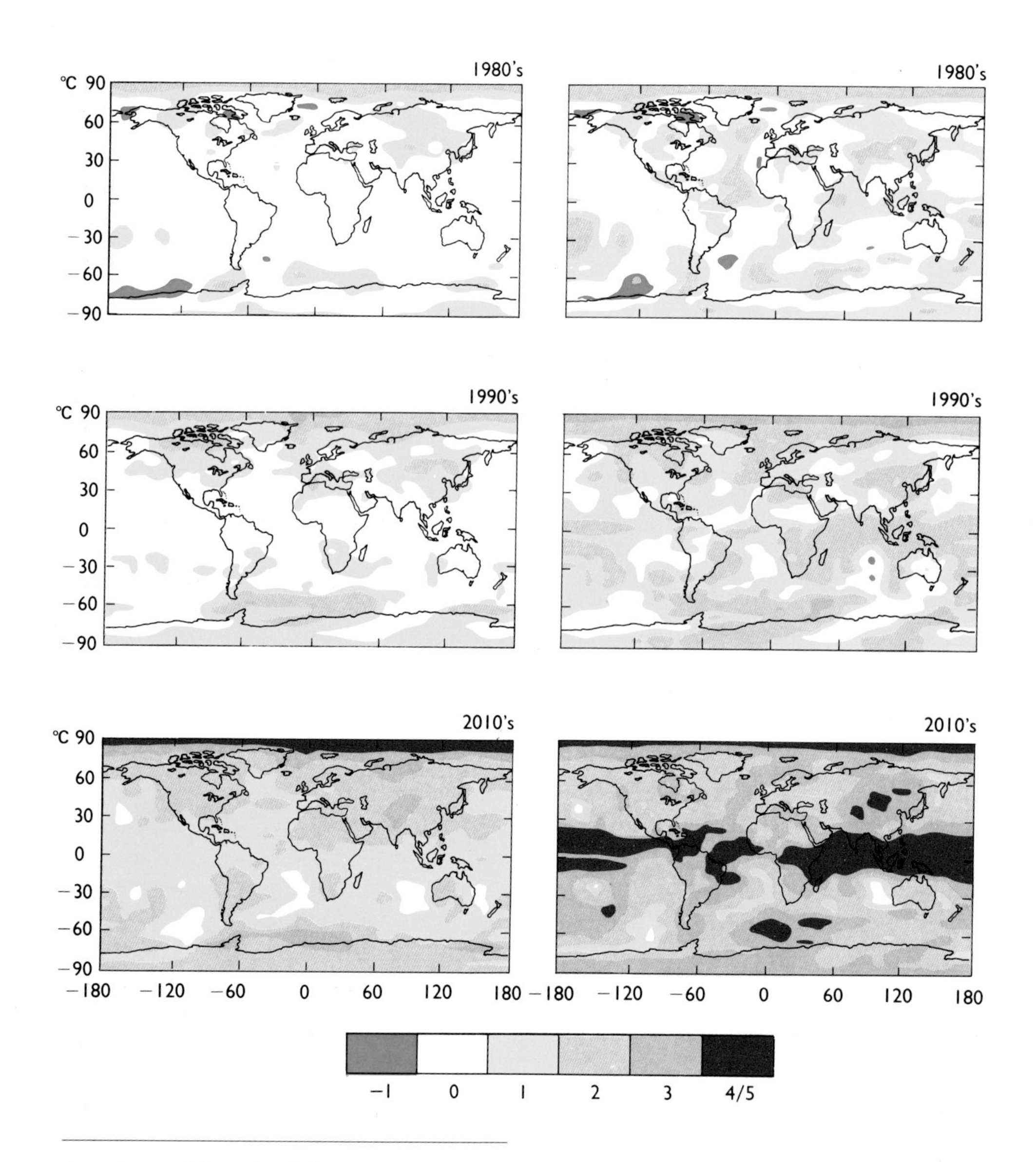

An estimate of the regional change in temperature as global warming progresses. The left-hand column shows the predicted difference in temperature between the 1950s and the 1980s, the 1990s and the 2010s. The right-hand column shows a measure of how significant these changes are
Source: J Hansen and others

drawn to the possibility of global warming as long as the 1890s, but then the concern was only about the burning of coal. There was no way account could be taken of the effects of CFCs as they had not yet been invented. It is only in the past ten years that scientists have realized that CFCs, methane and nitrous oxide are as important, in total, as carbon dioxide. And there may be other surprises in store. Even the change in climate will change the way that climate will change. As the world warms, people will use energy in different ways – for example, less heating in winter but more air conditioning in summer – and this will affect emissions of the greenhouse gases.

No scientist tries to predict what the weather of the world 'will be' in, say, the year 2030. Instead, they make assumptions about the way key variables will alter – population, energy consumption, and so on – and use GCMs to paint a picture of what the world *might* be like if these assumptions turn out to be good guesses. Often, a series of experiments will be run on the same computer model, with the basic assumptions varied to test how sensitive the results are. Experiments based on similar assumptions will be run on different models, to test how sensitive the results are to the variations between these models. The resulting collection of climatic portraits, similar in some respects and different in others, are not images of *the* future. They are images of *possible* futures, and are often referred to as 'scenarios'. By studying different scenarios from the model 'gallery', and looking at the assumptions on which they are based, we can decide what kind of world we would like to live in, and see what action we would have to take in order to make this future our reality.

Steve Schneider, climate modeller

We can appreciate the value of this approach by looking at a couple of examples resulting from the SCOPE review mentioned earlier. The SCOPE team began by considering scenarios for future energy use, deforestation (and other factors) to define how fast the composition of the atmosphere would change. Specifically, they wanted to know how long it might take for the concentration of carbon dioxide in the air to reach twice its preindustrial level of around 280 parts per million. This point is often used as a benchmark in assessing the rate of change in atmospheric composition and climate.

At the high end of the range, they concluded that, if no efforts are made to reduce fossil fuel consumption and industrial growth proceeds unchecked, this doubling would occur by some time around the middle of the twenty-first century. At the other extreme, if strenuous efforts were made to reduce carbon dioxide emissions, doubling could be held at bay throughout the twenty-first century and well into the twenty-second century. The picture changed dramatically when the SCOPE team added in the effects of the other greenhouse gases such as the CFCs, methane and nitrous oxide, measured in terms of their equivalent amount of carbon dioxide. In the high emission 'business as usual' scenario, the doubling date was brought forward to 2030, within 40 years or so. And even in the low emission scenario, the doubling date was brought forward to 2050, similar to the high emission scenario without the additional heat-trapping gases.

This exercise demonstrates two points. First, it highlights the importance of the 'other' greenhouse gases, neglected until recent years. Second, it emphasizes the substantial role that controlling emissions of the greenhouse gases could play in slowing down the rate of change in atmospheric composition. We *can* curb global warming, if we so choose.

The SCOPE investigators also reviewed estimates from equilibrium model experiments of how much this doubling would change global temperature. Based on all available model experiments, they considered that doubling should lead to a temperature rise of somewhere between 1.5 and 5.5°C above the pre-industrial level – a further 1 to 5°C above the present-day. (This rise in temperature would lag some time behind the buildup of greenhouse gases owing to the delay in ocean warming discussed earlier.) Worryingly, the latest and most complete model calculations all suggest warming at the higher end of the range of uncertainty. The SCOPE team concluded that global warming would be accompanied by a rise in sea level of somewhere between 20 and 165 cm as glaciers melt and the oceans expand.

This represents a substantial change in our planetary environment. Irving Mintzer of the World Resources Institute in Washington DC underlines the point: 'It is vital to understand what these changes could mean ... To find conditions like those projected for the next century, we must go back millions of years ... If the greenhouse effect turns out to be as great as predicted by today's climate models, and if current emission trends continue, our world will soon differ radically from anything in human experience.' The temperature projections represent an unprecedented change in climate, even towards the lower end of the range of forecasts. While there may be uncertainties still to be resolved, a more precise prediction will probably not be available for many years, awaiting improvements in our understanding of the climate system, more complex climate models and more powerful computers.

THE POLEWARD SHIFT

As the greenhouse effect takes a grip, the geography book definition of climate zones will have to be revised polewards, away from the equator. Climatic boundaries – equatorial, subtropical, temperate – will be relocated. Professor Pierre Morel of the World Meteorological Organization considers that 'it is very unlikely that global warming will produce of a new kind of climate. It is a question of climates being in different places.'

The temperature rise will be least at the equator – a zone of relative climatic stability – and greatest at the poles. As we have seen (on page 4) in the NASA computer simulations, the higher latitudes, those nearest the poles, are set to warm fastest and to warm most. There is no mystery about this. It is because of feedback effects caused by changes in the amount of snow and ice at high latitudes. In regions where the surface of the Earth, the land or the sea, is covered by snow or ice, it is reflective and a lot of the heat coming in from the Sun is bounced back into space without warming the surface at all. The bright snow or ice cover is lost as temperatures rise and the dark surface of the land or sea is exposed, absorbing more energy and amplifying the initial warming.

That's not all. The polar oceans contain heat trapped in the form of warm water beneath the ice, and this will be released into the atmosphere once the ice cover disappears, again boosting the

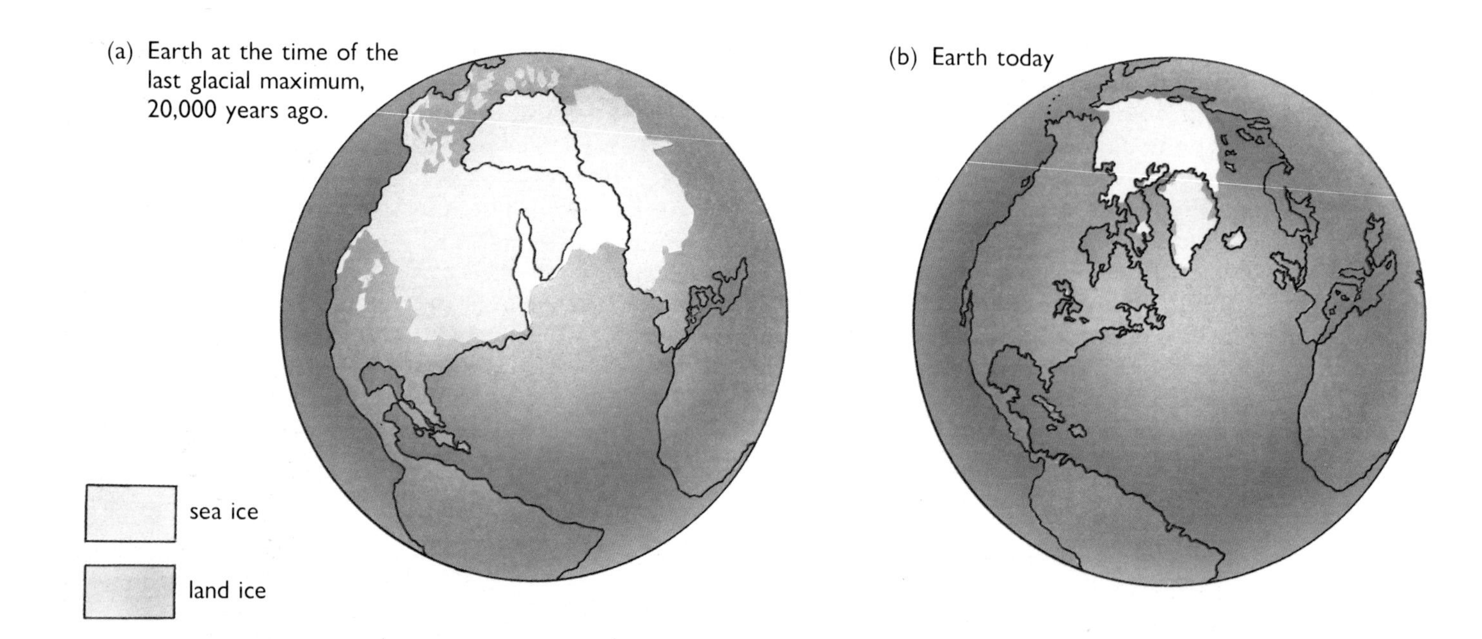

The ice masses of the Earth at the height of the latest ice age some 20,000 years ago (left) *and in the present day* (right)*. Global warming represents a similar change in the planetary environment, but taking place in decades rather than thousands of years. The Arctic Ocean may lose its ice cover altogether. Can polar bears tread water?*

Source: J Imbrie and K P Imbrie

original rise in temperature. As a result of these feedback processes, the greenhouse effect will be felt most strongly during the long polar winter. A global rise in temperature of, say, 3°C could be accompanied by warming of over 10°C in the Arctic winter. Once much of the snow and ice has gone – and this might happen if we let global warming progress too far – then these effects will no longer be important. It is to be hoped we never reach that stage. And there is yet another twist. As the permafrost – the frozen tundra of polar regions – melts, methane that is now trapped beneath the frozen surface will be released. This will accelerate the change in atmospheric composition, and the change in climate. Just one more example of the way in which the impact of global warming is intensified as the complex balance of nature is broken.

Temperatures in middle latitudes will change by roughly the same amount as the global average. To get some idea of what this might mean, you don't need a multimillion dollar computer, just take a look a few hundred kilometres equatorwards. For example, the climate of southern England might become rather like that of southwest France by the middle of the twenty-first century, with the southeast acquiring a drier semi-Mediterranean flavour.

Temperature is not all that will change. Global warming will also alter patterns of rain and snowfall. As the world warms, more moisture will be evaporated from the oceans and released into the air as snow and ice melts. Averaged over the whole world, this means that there will be more rainfall as the moisture is returned to the oceans. The so-called hydrological cycle will become stronger. But the patterns of rain and snowfall could be quite different from the ones we are used to. As the atmospheric circulation – the pattern of prevailing winds, of rain-bringing storm tracks, of dry anticyclones – varies, some regions and some times of the year may well become wetter, while others become drier.

One of the main driving forces behind the atmospheric circulation, the movement of wind and weather, is the contrast in temperature between the equator – where most energy is received – and the poles. Atmospheric circulation constantly strives to correct this imbalance in energy: air rises from the warm equatorial regions and moves out towards the poles carrying heat, exporting energy. As it travels polewards, it cools and sinks, returning eventually to

the equator. However, the energy that it carried is transferred further towards the pole by other wind systems, such as middle latitude depressions or cyclones ('lows'). The oceans also play their part in moving energy from equator to pole: warm ocean currents carry heat towards the poles away from the equator while cold ocean currents flow equatorwards, all playing their part in redressing the temperature imbalance.

This atmospheric flow is also strongly affected by the rotation of the Earth, which deflects the wind currents as they travel polewards, producing a strong stream of westerly winds in middle latitudes. Mountain ranges further divert the flow, while temperature differences between land and ocean play an important role in shaping the character of regional wind circulations. The most notable example of this latter effect is in the monsoonal regions where the seasonal heating of the continental landmasses produces a marked change in airflow – and rainfall – from season to season.

The greenhouse effect is a disturbance of the energy balance. More energy is trapped near the Earth's surface and less at higher levels in the atmosphere. This alters the flow of air in the lower atmosphere. But global warming will change the energy balance in other, more subtle ways. In particular, the temperature difference between equator and pole will be reduced – and the atmospheric circulation will change to reflect this. All aspects of the weather will be affected, and the resulting changes will be superimposed on the poleward shift of climatic zones that is the basic effect of global warming.

Will this mean more extreme weather events? Or fewer, because the weather machine has less work to do as the poles warm? It is impossible to say at this time. But, as climate moves beyond its present-day bounds, we will certainly experience what appears to be a greater frequency of unusual events as we slowly become accustomed to a new, and ever-changing, set of climatic conditions while what was once considered extreme becomes the new norm. And this is the real significance of events such as the recent heatwaves in the Mediterranean, the flooding in Bangladesh, the severe hurricanes in the Caribbean and the North American drought of 1988. We may never know whether or not these particular events were products of the greenhouse effect; but they do provide a graphic foretaste of what might be in store in the very near future – as the polar warming of recent decades spreads towards the equator during the 1990s, disturbing atmospheric circulation and regional climates world-wide.

THE BREADBASKET DEPLETED

Jim Hansen of NASA attracted considerable attention during the summer of 1988 when he testified before the US Congress that the drought then affect-

'We've promoted very aggressive high-tech farming over the last fifteen years. We have pumped our water tables down to extremely low levels. We have stripped the surface of a great deal of vegetation that would protect the soil from wind erosion and we've also stripped our land of quite a bit of the wetlands which would help replenish the water table when it finally does rain. So we would be in exactly the worst position if we're going into a long, dry greenhouse period in North America.'

Hugh Ulrich, futures analyst

Summer 1988: (above) *Ron Rivinius and his failing sunflower crop and* (below) *Ralph Waldron highlights the impact on grain prices*

ing many parts of the Midwest might indeed be the result of the greenhouse effect. This wasn't outright speculation. It was based on one of the clearest predictions from climate model experiments.

As temperatures rise, the northern continental interiors – North America, Europe and Asia – are likely to dry out. There are two reasons for this. First, higher temperatures will mean less water stored as snow and a drier start to the growing season in spring. Second, higher temperatures will mean more evaporation and more water lost to the atmosphere throughout the summer half-year. In some areas, this may be compounded by decreased rainfall. In others, it might be offset by greater rainfall amounts.

Drier conditions, drought, in North America, Europe and parts of the USSR will have an extremely adverse effect on traditional crops. Even with no change in rainfall, a rise in temperature of only 1°C could reduce cereal yields by ten percent in regions such as the American Midwest. With less moisture in the soil, and a temperature rise of 3 or 4 °C, substantial reductions in harvests are inevitable. And harvest failure in North America is a problem for the whole

world. The region provides 90 percent of the world grain surplus – the food that is not needed by the country that grows it and is available for trade or aid. If yields decline, this surplus will be lost. And if the surplus goes, prices will inevitably rise.

This particular scenario for the future is one of the most threatening of all the global warming predictions. How will the poorer nations of the world cope when, ravaged by climate stress, they find themselves unable to buy food on the world market, while the flow of aid is stemmed as donor nations protect their own supplies? It is difficult to avoid the conclusion that mass starvation will result.

WHEN THE MONSOONS FAIL

While the rise in temperature will not be as marked at lower latitudes as at the poles, this does not mean that tropical and subtropical latitudes will escape the heat trap. Indeed, the climatic stability of these regions means that even a minor change will have great repercussions. Unused to climatic change, both natural ecosystems and humanity will find it difficult to adapt to a new set of climatic conditions. The greatest consequences are likely to stem from changes in water availability, a critical factor throughout equatorial regions. Changes in the distribution of rainfall from one region to another and from one season to the next will have tremendous implications for agriculture. In some cases, the net result will be beneficial – in others, it will be adverse. Much will depend on how well prepared farmers are to take advantage of favourable developments and to combat those that are not.

The fate of the monsoon may well be the most significant aspect of global warming for southern Asia and parts of Africa. In these regions, the monsoon is responsible for much of the annual rainfall. In winter, the land becomes cold, and cools the air above it. The cold air sinks downward and flows as a steady wind out to the sea. As summer approaches, the land warms faster than the ocean and, as soon as the temperature of the land exceeds that of the sea, the wind reverses. Now, hot air rising over the land draws in moisture-laden air off the sea. As that air rises over the land, it sheds its burden of moisture as rainfall. Obviously, any factor that changes the temperature balance between land and ocean, as global warming surely will, is likely to affect the strength – the timing, intensity or duration – of the monsoon circulations of lower latitudes. Computer models suggest that the greenhouse effect will intensify the monsoon in parts of southern Asia but that the monsoon will become weaker to the north. Essentially, the flow will not penetrate as far inland as it does today.

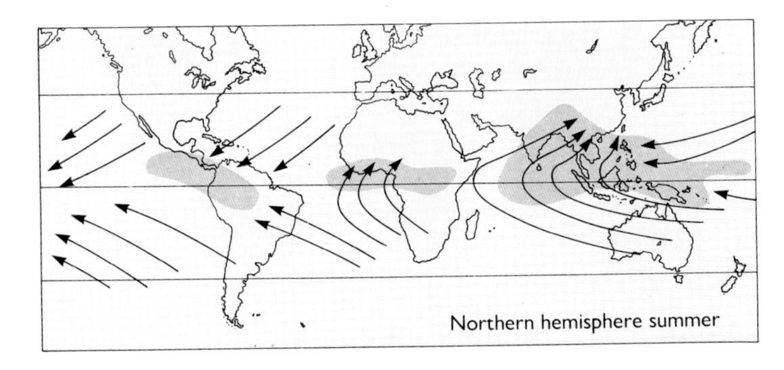

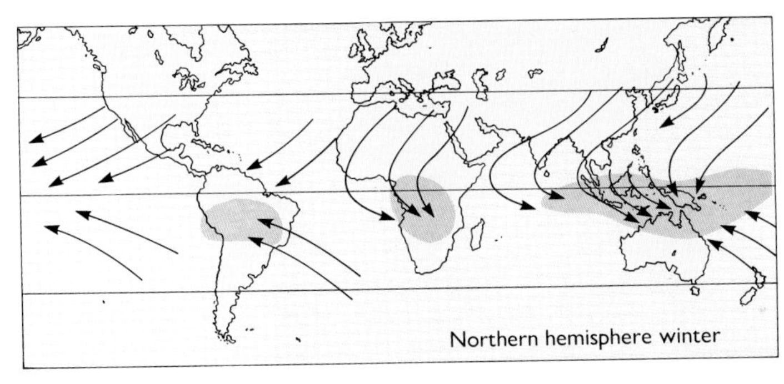

The monsoons of Africa and Asia in the Northern hemisphere, summer (upper) *and winter* (lower). *The areas experiencing the greatest seasonal rainfall are shaded and the most important wind systems are indicated by arrows. The wind reversals associated with the African and Asian monsoons are clear*
Source P J Webster

There is another major atmospheric process which will be affected by global warming and which exerts its influence throughout the tropics and subtropics. This is known as the 'El Niño-Southern Oscillation' phenomenon – ENSO for short. El Niño, the Christ-child, is a warming of the tropical Pacific Ocean near South America. Occurring every few years, it is associated with a major disturbance in the ocean current systems of the Pacific and with a shift in the overlying atmospheric circulation, the Southern Oscillation. ENSO produces a major upheaval in the climates of many regions, particularly in the Southern Hemisphere.

One of the strongest ENSO events this century occurred during 1982 and 1983. In South America, northern Peru, Ecuador and southern Brazil were hit by torrential rains causing landslides and flooding. Peru faced losses of $1 billion as the potato crop was ruined. Two hundred thousand square kilometres of land were flooded in Argentina. Elsewhere, drought was the dominant hazard as rainfall patterns shifted across the Southern Hemisphere. Southern Africa suffered a catastrophic drought. Seventy-five percent of the livestock in the Bantu 'homelands' had to be slaughtered and South Africa, usually an exporter of maize, had to import 1½ million tonnes that year. Indonesia suffered a cholera epidemic as drought generated famine and eastern Australia experienced the worst drought in living memory.

Will such events become more frequent, more intense, as global warming develops? Again, we cannot say for sure – although we do know that there is a strong link between ENSO events and global temperature. The warming of the tropical Pacific during an ENSO event will reinforce the more general warming produced by the greenhouse effect. And warming of the tropical Pacific caused by the greenhouse effect may boost the intensity of ENSO events. It may not be a coincidence that the most intense ENSO of the century has occurred during the 1980s with global temperature at a record high.

A DROWNING WORLD

Sea level rise is the aspect of global warming that is likely to have the greatest impact. About half the world's population live in low-lying areas. Sea level has already risen by over 10 cm this

Flooding in Bangladesh

century as global temperature has increased by 0.5°C. This rise in sea level can be explained almost entirely by the expansion of sea water as the oceans have warmed. Each further 0.5°C rise will cause an equivalent expansion of the seas – but, over decades to come, there will be more water flooding into the sea from melting mountain glaciers. Over the next 40 years, sea level could rise by as much as 20 cm a decade – 20 times as fast as over the past 100 years. By the end of the 21st century, sea level could stand 2 m higher than at present.

'The global climate problem is fundamentally interdisciplinary in character. It cuts across the boundaries of the traditional realms of chemistry, physics, biology and politics. One has to understand the complex interactions not just of the biogeophysical systems but also economic and political systems which determine investment choices and the policy strategies of national governments. In addition, it's fundamentally international in character, which means that any strategy which will reduce the risks will have to engage the cooperation of people in a large number of countries, both industrialized and developing.'

Irving Mintzer, World Resources Institute

By the year 2100, much of the summer pack ice in the Arctic Ocean may well have disappeared, but this will not add to the problem of sea level rise, as polar ice floats. Like the ice melting in a cocktail, polar ice simply displaces its own volume of water: its disappearance will not cause the overall level of the ocean to rise. At the other pole, though, there is concern that part of the great Antarctic ice sheet may become unstable and break off. And as this ice is grounded on land, a substantial rise in sea level would result – of, perhaps, 5 to 6 metres. This threat is, however, remote in time. It would take a rise in temperature many degrees greater than we are likely to experience in our lifetime to trigger such a catastrophic event.

Even today, however, coastal flooding is a major cause of loss of life and livelihood, especially in the Third World. In Bangladesh, then East Pakistan, 300,000 people lost their lives in one flood, the result of a storm surge, in 1970. In 1953, flooding due to a severe storm in the North Sea took the lives of nearly 2000 Dutch people and inundated 150,000 hectares of agricultural land. In the wake of the 1953 disaster, the Dutch built massive new protection schemes. The Bangladeshi government, with limited resources, could do little more than wait for catastrophe to strike again.

Three billion people live in low-lying coastal regions. Many major cities – London, Venice, Miami, Shanghai, Leningrad, Rio de Janeiro, to name but a few – are located close to sea level. Docks, roads, railways, airports, as well as industry and homes, will be at risk. Many nuclear power stations are built in coastal zones, taking advantage of copious water supplies, and these too will be vulnerable to flooding. Sea level rise would also wipe out many valuable coastal ecosystems – the mangrove swamps of Florida and Bangkok, the marshes of Essex in the UK.

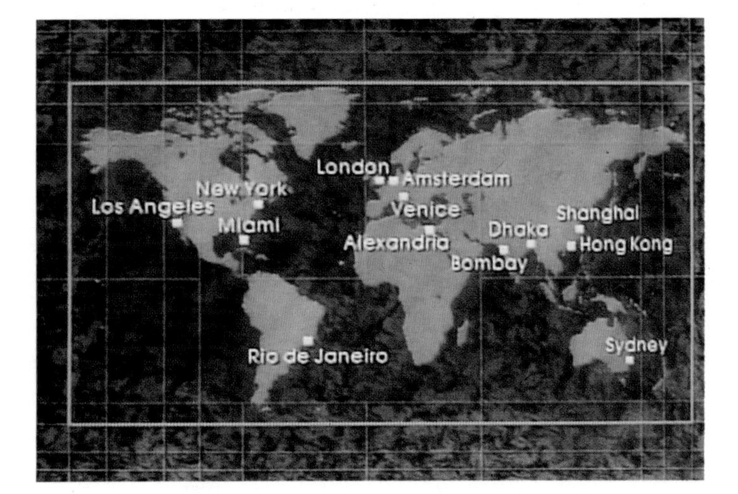

Cities at risk as sea level rises

Many of the people of the Third World have moved to vulnerable coastal plains and deltas where fertile agricultural land is found. A rise in sea level of 2 m would displace more than a fifth of the Egyptian population, living on the Nile Delta. It would affect a third of the Bangladeshi population whose homes are on the Ganges Delta. People in neighbouring areas would face a greater risk of storm-induced flooding. What can be done to protect these regions as the frequency of flooding increases? Mass evacuation may be the only solution – but where will the people go?

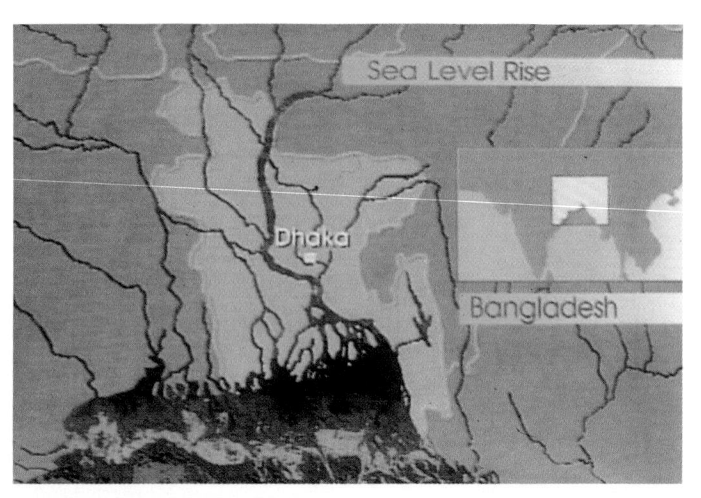

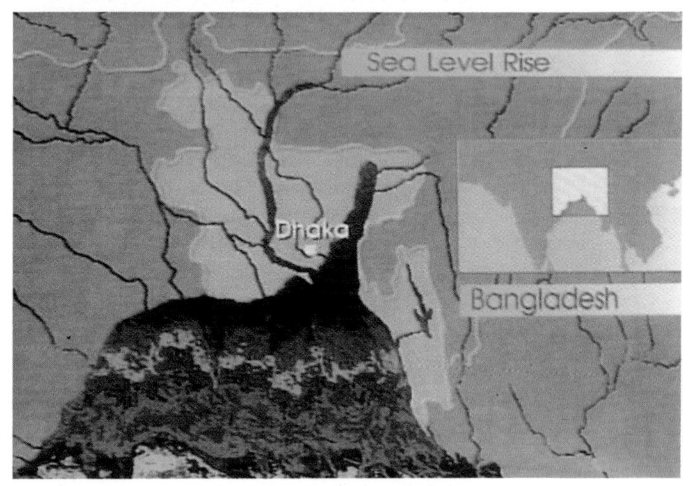

Bangladesh today (above) and after a 3m sea level rise (below)

The problem of sea level rise is compounded by the fact that it takes a long time to plan and build coastal defences. Few, if any, existing schemes have taken the possibility of accelerated sea-level rise into account. The Dutch coastal defences, improved after the 1953 floods, were built to withstand anything, short of the kind of event that 'ought' to occur once every 10,000 years. But a rise in sea level of just 1 m would increase the probability that the barriers will be breached to once every 100 years. If the system is not strengthened, the odds are that it will fail by the end of the twenty-first century. London's Thames Barrier is also inadequate to meet the threat posed by rising sea level. In evidence to a public inquiry into plans to build a new nuclear power station at Hinkley Point in March 1989, David Fisk, Chief Scientist in the UK Department of the Environment, noted that the Barrier's design only allows for a rise in sea level of 30 cm over the next century. This rise could well occur within a couple of decades.

It has been estimated that improving the UK's coastal defences alone would cost an astronomical £10 billion simply to counter the rise in sea level over the next 60 years. World-wide, the cost could run to many thousands of billions. While the richer nations of the industrialized world might find the funds to invest in further protection, the chances are that the Third World will not. Faced by more immediate problems, it is unlikely that resources could be freed

The Thames Barrier

without jeopardizing survival chances in the present day. The need for aid and assistance is clear.

INTO THE TWENTY-FIRST CENTURY

What trends in climate might occur over the next 30 years? As we have seen, it is impossible to make firm predictions. There are too many unknowns, too many uncertainties. But we can speculate in an informed way. We can construct a scenario for coming decades which is consistent with what is understood of the possible course of global warming over that period. And *possible* is the key word. In many ways, a good scenario is one which does not 'come true'. By warning us of dangers ahead, a scenario may encourage us to mend our ways, ensuring that the bad times are not so bad after all. Like all scenarios, this future chronology serves to highlight key issues, to point out potential consequences, rather than as a definite forecast. It is extreme. But it is well within the bounds of possibility . . .

During the early 1990s, the warming that has previously been largely confined to the sparsely-populated polar regions of Alaska and Siberia spreads south. Scandinavia and northern European Russia experience a sequence of mild winters. There are many benefits – lower heating bills, greater industrial output, fewer transport problems – but there are also disadvantages. Less snowfall in the southern parts of these regions reduces the water available during the early months of the growing season and agriculture suffers accordingly. Pests that used to be killed off by cold winter weather thrive. But, all in all, the inhabitants of these areas consider it to be a change for the better.

Elsewhere in the world, a major El Niño event, surpassing the record-breaker of 1982–83, brings climatic disaster to many parts of the Southern Hemisphere. Flooding in South America, prolonged drought in parts of southern Africa and Australia and partial failure of the southeast monsoon lead to widespread harvest failure. A massive aid effort is underway but food reserves are in short supply. North American production was hard hit by the drought of 1992.

'When I turn on pieces of equipment in my home, it causes electricity to be generated by a power plant in Maryland, Virginia. The emissions that result from that coal combustion will affect the climate not only over Washington but over India, over Burma and over China. It will contribute to the climate change that delays the monsoon and reduces yields all over southeast Asia, affecting the life of a rice farmer in Bangkok. By the same token, the flooding of his paddy to grow that rice will release methane into the atmosphere and that methane will itself add to the risk of future global warming.'

Irving Mintzer, World Resources Institute

By 1996, the effects of global warming have spread throughout northern middle latitudes. Drought has affected the US in three of the past five years, the USSR twice, and Canada, though benefiting from warmer weather, is also suffering water shortages. Harvest production, though, is not down substantially as yet and farmers in North America have had financial losses met by their respective governments. Nevertheless, the world grain market is showing signs of nervousness and the world price of wheat is edging upwards as reserves are depleted.

Following a record mild winter in North America, the summer of 1999 proves the crisis point. Drought affects the grain-growing regions of the US and the USSR for the third successive year, grain harvests drop by 60 percent and world reserves fall to a supply of only a few days. The US government bans all exports of grain, brings food aid programmes to an immediate halt, and the world price of wheat triples. The USSR stops providing its satellite nations with grain. Recurrent harvest failures and famines in the Third World go virtually unnoticed as the North concentrates on coping with the change in its own climate – and on belatedly attempting to control carbon dioxide emissions. Calls for assistance to the UN are to no avail.

As the flow of food aid dries up, mass starvation hits the African sub-continent as the endemic famine of the past three decades can no longer be contained. The population of the worst-affected regions drops to zero. Most self-sufficiency programmes have failed as climatic change has wreaked havoc with long-term planning. On the Indian sub-continent, the northern regions are also hard-pressed, although food shortages are being met by improved production in the south and regional aid programmes are proving effective.

In the autumn of 1999, the first cyclone of the season hits Bangladesh, drowning 650,000 people on the Ganges Delta. As the world watches horrified, a second cyclone, only weeks later, kills a further 80,000 people – and a mass exodus from the region begins. Frontiers are closed as refugee camps overflow and a UN peace-keeping force is brought in to reduce tensions between the neighbouring countries.

By the early years of the twenty-first century, mass migration has become commonplace. The southern border of the US has become heavily fortified in a forlorn attempt to stem the tide of refugees from Mexico and other Central American nations, now suffering continuous water shortages. Refugee camps are scattered throughout the European nations on the Mediterranean coasts and the northern European nations are being pressured to take in their share of the new wave of 'boat people'. In the Middle East, conflict over resources has led to the rekindling of old enmities and warfare breaks out. After two nuclear strikes, the US and the USSR intervene, demonstrating an unprecedented degree of cooperation in forcing an end to the hostilities.

The last remaining area of primary rainforest anywhere in the world, in a wild-life sanctuary in northwest Thailand, burns to the ground in 2004 in one of the worst forest fires in living memory. A victim of prolonged drought, it is an ironic end to one of planet Earth's most valuable ecosystems. With much of the world's forests felled by axe and chainsaw, the final blow was struck by global warming.

And sea level rises steadily.

By the year 2010, sea level stands some 30 cm higher than at present. Recurrent flooding has led to the abandonment of many coastal areas in the Third World and new defensive systems are being hastily erected in the North. The first major casualty in the industrialized world is Florida. The year is 2013. Hurricane-force winds generate a storm surge which inundates beach-front hotels, devastates the mangrove swamps, and destroys roads and bridges down the Florida Keys – over 1000 people die, over 100,000 are made homeless. The following year, disaster strikes London . . .

But this, remember, is a *scenario*, one that we do *not* wish to come true. Unfortunately, it is all too plausible, as we can see by looking at what has already happened.

3

'SECOND ONLY TO NUCLEAR WAR'

The winds of change that will blow across human society as we move into the twenty-first century are not completely without precedent. Extremes of climate have occurred in the past, and these events can give us some guidance as to what we might expect as such conditions become the norm. Our experience of real events – rather than the statistical predictions of computer models – makes the idea of future global warming come alive. And as we look further into what the future may hold in store, we can see why the Toronto Conference delegates referred to the threat as 'second only to nuclear war'.

In all the talk about droughts and floods, crop failures and socio-economic consequences, one fact sometimes gets obscured. Extreme weather can have a direct effect on human health and warmer, wetter conditions are good for organisms that carry disease and for many parasites. Problems such as malaria, eradicated from many temperate regions such as the US and Europe, could return as climate changes. And extreme heat is as powerful an assassin as extreme cold, as a study of three heatwaves in Los Angeles by Margaret White and Irva Hertz-Picciotto at the Lawrence Berkeley Laboratory in California demonstrates.

In 1939, a seven day heatwave hit the Los Angeles area, with peak temperatures reaching 42°C. On the fifth day of the heatwave, death rates reached nearly three times the normal rate. The rate in the over-85 age group was six times higher than normal. In 1955, another week-long heatwave struck Los Angeles, but this time the peak temperature edged above 43°C. Death rates rose to four or five times the normal rate over all age groups, eight times normal for the over-85s. By the time of the third event in the 1960s, improved healthcare and other factors had blunted the impact of extreme heat. Even so, mortality over all age groups reached nearly twice the normal rate during the heatwave of 1963 and it was two and a half times the expected rate in the over-85s.

In cities such as Los Angeles, global warming will be reinforced by the waste heat from homes and factories and the heating of concrete buildings, roads and so on – the urban heat island. The effects of rising temperature will be compounded by photochemical smog. It is certain that a lot of people, particularly the young and the elderly, will die simply because it is too hot. But urban heatwaves are a small-scale phenomenon. The weather and global climate have had an impact on a much grander scale.

IN THE WAKE OF THE FLOOD

John Pilger, journalist and author, describes how he heard about the terrible catastrophe that hit East Pakistan, now Bangladesh, in November 1970.

Early in the morning, a telegram arrives at the *Daily Mirror* from a local correspondent.

TELEX TELEX

FIVE LACKS GONE. DEFINITE. POSSIBLE ONE CRONE.
NO SOULS SIGHTED PATUAKALI. WAVE THIRTY FEET.
MINIMUM ONE CRONE CATTLE DONE FOR NOAKHILI.
TERRIBLE STATE AFFAIRS. RESPECTFULLY REQUEST
MOST URGENTEST IMMEDIATE ATTENTIONS GOVERNMENT
AND NEWSPAPER. TOPMOST PRIORITY. COME QUICK.
SORROWS. FAKHRUDDIN.

A lack is 100,000. A crone is one million. Forced by a tropical cyclone, a tidal wave had inundated the coastal regions of the Bay of Bengal killing between 500,000 and 1 million people. Between 20 and 30 feet high, the wave glowed, luminescent, in the darkness of midnight. Their houses drowned, the handful of survivors climbed the few trees not swept away – and huddled there for 2 hours before the waters began to subside. Then the winds came, at speeds of 100 miles an hour, tearing the frantic survivors from their perches. And slowly the waters receded, carrying away the bodies of the dead.

The great rivers of the Ganges and the Brahmaputra merge in the delta region of the Bay of Bengal. A network of islands, much of the land is flooded by high tides at least twice a year. But the soil is fertile. Rice is grown year-round and many cattle are raised. Nowhere else in the country can the population grow sufficient to feed themselves. As a result, more than a fifth of the Bangladeshi people live in the low-lying coastal plains – despite the risk. In the words of John Pilger, 'People have to live where no people should live, sheltering behind trees planted in a single elevated line across the middle of each island as a pitiful rampart against the storms. This is an extremely dangerous place: a geographical coffin.'

We cannot improve on John Pilger's account of what he found when he arrived in the disaster zone. There are few journalists who write with the understanding, compassion and commitment that he manages to share with the reader. There would be less complacency in the world, the world would be a better place, if there were more like him. From his book *Heroes* (Pan, 1986):

'The first island we approached was Hatya. About 2,000 people had lived on it, but now it appeared like a polished coin, save for one leaning house with a grey shadow on its southern rim. As we drew closer Kamal leapt to his feet.

'"People." he shouted. "All people."

'The shadow was dead people. Some of them were sitting upright; some were holding each other.

'"There was a primary school right here,' said Kamal. "Where is it?" Kamal anchored near the breakwater formed by the bodies, of which there were several hundreds. The headmaster surveyed his relief boxes, then picked up the lime and trudged into the sludge. We followed, going calf-deep through sand and silt that looked as though it had been swept by a giant rake. We halted where a wedge of children's bodies lay. The headmaster threw lime on them, cursed and threw the box to the side. It seemed such a perfunctory, inhuman act, but it was not; tears streaked his face.

'The house on its own was a monument to the impossible; the reason why it had not been swept away with everything else was soon evident. There may have been fifty bodies inside; heavens knows how many. They were people who had heard the roar and had huddled together in the most secure house, and somehow their combined weight had kept it upright. In the branches of the tree behind there was the body of a child.

'A man who was an organiser of the Awami People's League, the Bengali nationalist party of East Pakistan, told me angrily about an emergency evacuation plan which had not been implemented and a radio warning that had spoken only of severe winds and protective embankments, which the government in Islamabad had promised to build but had not. "We are not animals," he said. "There is no reason why we should be the objects of God's hatred year after year."

'He led me to a woman called Nurunnessa, whose madness, he indicated, would illustrate what he was saying. For five years, during five natural calamities, she had watched her immediate family decline from six members to one – herself. In May 1965 her oldest son, Oli, had been drowned in a flash flood. In December of the same year her second son, Khan, had been one of 10,000 killed in a cyclone. In October the following year her mother-in-law had died from injuries after her six-year-old daughter, Moli, had been swept away while on an errand. During the harvest, he said, it was common to work around the clock and her husband, Seraj, had been in the paddies at midnight on Black Friday, the 15th, and he was still missing.'

Aid was slow in coming. The Pakistani government in the western province held back, believing that the tragedy would hamper the efforts of the Bengali nationalists. The western nations, reluctant to offend the Pakistani government who they regarded as a bulwark against China, held back. By the end of the first week, no aid had reached the worst-affected areas and people were dying at a rate of 20,000 a day – from cholera, from dehydration and from their injuries. As Pilger observes, 'The aftermath of the 1970 cyclone told us much about the nature of Western aid and its purpose: preserving in power a political and economic order incompatible with the interests of those in greatest need of genuine assistance.'

The world has progressed since the early 1970s. Or has it? September 1988, nearly two decades later. Another flood hits Bangladesh – this time the result of heavy rainfall – leaving 20 million people homeless, 1000 dead and 2 million tons of food destroyed. In the aftermath, Chris Patten, Britain's Minister for Overseas Development, commented: 'We have, all of us in the international community, been looking for some years at longer term ways of coping with disasters. We are going to have to look more urgently at what can be done.' If the catastrophe of 1970 was not sufficient to galvanize the international community, to force them to look harder in the search for solutions, what will it take? Andy Rutherford of War on Want hit the nail on the head: 'This is one year to the day since the last floods in Bangladesh but lessons are not being learned.'

How many years, how many tragedies, does it take for the international community to learn? If we cannot protect vulnerable people today, how will they survive tomorrow the greater hazard of global warming? When disaster strikes, when harvests fail, the people who suffer most are always the poor. To quote James Grant of UNICEF, speaking in 1980: 'Some 15 million small children die each year.... They die very quietly; one hears very little about them; they come from the world's poorest families, who themselves are the weakest and the most powerless members of those powerless families.' We allow this to happen today. Will we condone even greater outrages as global warming adds to the unbearable stress already being experienced by many of our fellow inhabitants of this planet?

WHEN THE BREAD-BASKET BECOMES A DUST BOWL

If not death by drowning, then through starvation. 1972 sets the scene. In India, the monsoon failed and brought an eight percent drop in rice production. Off Peru, the anchovy fishery failed. There were droughts in Australia and south America. The Sahel drought intensified. People went hungry in many parts of the world.

Many, if not all, of these events were related to the onset of El Niño in the eastern Pacific Ocean. Off the west coast of South America, a broad tongue of cold water welling up from the deep carries nutrients to the surface, providing an abundant supply of food for the anchovy and other fish. When El Niño warms the eastern Pacific, this tongue of cold, nutrient-rich water shrinks, concentrating the shoals of fish into a narrow region of the ocean. This is good news for the first fishing boats to locate this area – which is why this gift of the oceans, occurring every few years in November and December, has been given the name 'The Christ-Child', El Niño. But in 1972, the gift was short-lived. Overfishing in previous years had reduced the stocks of anchovy and in 1972 and 1973 the fish simply disappeared from the usual grounds. Until that time, Peru had been the leading nation in the world in terms of tons of fish caught each year. In 1972, it fell to fourth place in the league table.

Soviet food production was also low – 12 percent short of expectations – and the region around Moscow experienced its worst drought for 300 years. A lack of winter snow ruined the crops: there was no blanket of snow to protect the young plants from frost and less moisture from melting snow in the spring. The North American harvest was also down, hit by floods in the Midwest.

Overall, world food production fell by two percent. Not a large shortfall, but it was the first time since the late 1940s that production had failed to increase from one year to the next. Encouraged by the US government, keen to further the process of détente, the USSR bought up one quarter of the US wheat crop that year, together with grain from Canada, making a total of 28 million tons. The purchases started early in the season and were unpublicized. These secret deals, combined with the decline in world food production, led to panic on the markets – hoarding and speculation forced prices up to four times their previous level.

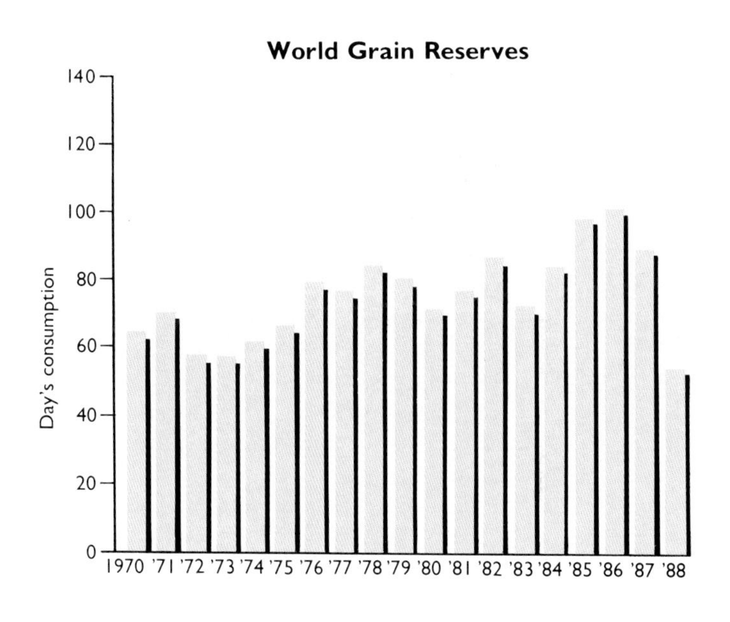

World grain reserves Source: US Department of Agriculture/ Worldwatch Institute

The world's reserve of grain, previously enough for 60 days supply, fell to 20 days supply. Two months' reserve is precarious enough, but as a result of the 1972 panic the entire world was just three weeks away from starvation. And many people did starve, especially in India where the harvests had failed but there weren't sufficient funds available to purchase alternative supplies on the inflated world market. All this because global food production had fallen by just two percent. Once again, feedback is at work. The impact of this relatively small shortfall was amplified, exaggerated, by the way in which the world's economic system – in this case, the grain market – works.

Worldwide, production improved in 1973, and again in 1974. But in the latter year, the grain-growing regions of North America were badly hit by drought. Since the great agricultural successes of the 1960s, many people had been saying that modern technology had solved the problem of climatic variability, at least in a nation as developed and mechanized as the US. Technology could 'climate-proof' food production. 1974 gave the lie to those claims. Corn yields fell by a fifth as a result of drought in the Midwest in spite of everything that modern technology had to offer. The problem recurred in 1988 when the worst drought for many years hit the American Midwest. Farmers who had made huge financial investments in modern machinery saw their tractors and combines standing idly in their sheds while their crops withered in the fields.

By 1986, world food reserves had climbed to a record high of 100 days supply. But harvest failures in India in 1987 had cut back reserves from that high before the climatic screw turned again in 1988. Concern about the fate of the United States harvest resulted in 'a tremendous increase in the volatility of the market', according to Ralph Waldron, futures analyst with Stotner and Company, Chicago Board of Trade. 'In a very short time, a matter of probably eight to ten weeks at the most, the market rallied from the seven dollar level to the eleven dollar level.' The drought of 1988 saw world food stocks reduced to their lowest level since 1972 – and it was followed by a winter with little snow, causing the same problems that the Soviets had experienced in 1972. The mild winter encouraged seeds to germinate early but it was followed by blasts of cold Arctic air in February, causing more damage. And in March 1989, 60 mile an hour winds swept dust clouds hundreds of miles wide across the Midwest, thick enough to close the interstate highway in Kansas because of bad visibility and to conjure up images of the Dust Bowl of the 1930s. Dust storms in winter, when there should be at least some moisture in the soil to hold down the dirt, are a very serious indicator of drought.

In 1988, the US consumed more grain than it produced for the first time since the Declaration of Independence. This is an ominous portent for the 100 or so countries around the world that depend on American production for a major proportion of their food. A series of crop failures – perhaps just one more – could swiftly result in crisis. In March 1989, John Young of the Worldwatch Institute in Washington DC pointed out that 'we have been accustomed to two droughts a decade somewhere in the world. If we are now moving to a temperature range where we can expect three droughts a decade, we have to rethink the adequate levels of stock. . . . If this year is going to be bad, the crisis could come as soon as May for the wheat crop, July for the corn crop.'

As we write, it is April 1989. By the time you read these words, you will know if the crisis did hit in May or July, or whether heavy spring rains have saved the US from two agricultural disasters in successive years. Whatever the outcome this year, loss of production in the heartland of the US poses one of the most immediate threats associated with global warming. Catastrophe on a global scale could be triggered within the very near future. James Hansen of NASA is forecasting *six* droughts in the US during the 1990s, not just three.

Perhaps in years to come, the US will be forced to buy wheat from the USSR, their production boosted by the new agricultural lands of the north. Whatever happens, the superpowers will ensure their own survival, trading food in accordance with the laws of supply and demand. As food becomes shorter and shorter in supply, prices will rise and only the rich will be able to afford to eat. Those with no financial muscle – the poor of the Third World – will die, in their millions, as climate stress hits the tropics and subtropics, as the famine we have seen in the Sahel and Ethiopia extends into more and more areas.

'One child dies every four minutes in my country.'
Glaca Machul, Minister of Education, Mozambique

DEATH IN THE SAHEL

Famine in the Sahel region of Africa has become the archetypal image of climatic disaster over the past 20 years. The region that has been affected is, strictly speaking, larger than the region that properly justifies the name 'Sahel', the area of Africa just to the south of the Sahara desert extending across six African states: Mauritania, Senegal, Mali, Upper Volta, Niger and Chad. The drought-affected region extends further south, and also further east to encompass Sudan and Ethiopia, making a band across Africa roughly in the latitude zone from 10° to 20°North. Many of the most tragic pictures of the consequences of recent droughts have, in fact, come from Ethiopia and Sudan. Droughts in this larger Sahel zone have killed more people in the past two decades than the wars and guerrilla activity that also made the news.

The crisis began in the late 1960s as climate changed for the worse. The shock was considerable because it came just after a time of relative plenty. Rain-

fall in the Sahel region reached a twentieth century peak in the 1950s. In the four decades up to the end of the 1960s, the human population of the region increased by a third while the number of livestock doubled. These changes were hastened by aid from the developed world – aid that increasingly transformed an essentially nomadic lifestyle based upon herding animals into a more settled, farming way of life. As soon as the weather turned for the worse, that new lifestyle proved completely inappropriate for the region.

When the rains failed in 1968, the drought that set in was, in itself, no worse than many droughts that the region had suffered before. But never before had there been so many people to feed. And never before had there been so many animals, eating the grass and then stripping the leaves from the trees in their hunger. Never before had so much of the land been under the plough, so that exposed soil could dry out and blow away on the wind when the rains did not arrive. And, unlike in previous droughts, the rains stayed away – year after year. Since 1968, the rains have never returned to the levels that were regarded as 'normal' earlier in the 1960s. A temporary recovery in 1974 and 1975 took the Sahel rainfall back to 90 percent of the old normal, but it then fell to new lows in the 1980s before recovering, in 1988, to the 'high' of the mid-1970s. What used to be the normal rainfall in the Sahel now seems to be the occasional, increasingly rare, exception – drought has become the new norm.

In the first phase of the Sahelian disaster, in the five years following 1968, more than 100,000 people died in the Sahel proper. Many more than this died further east, in the Sudan and Ethiopia, but official figures for the toll have never been released by those countries. In misguided efforts to recover from the disaster, even more land was put under the plough and cultivated. But the average yield of grain fell as the poor soil was rapidly exhausted. The nomads knew that they could only extract so much from the soils of the region. If this were not the case, farming would have been adopted centuries earlier. In Niger, the yield used to be 500 kilos for each hectare of cultivated land in 1920. In 1978, the figure was 350 kilos.

Sahel rainfall during the May–October season. Because rainfall amounts vary widely from place to place and from season to season, a simple index of rainfall amount which takes these factors into account has been used. The units are arbitrary Source: G Farmer

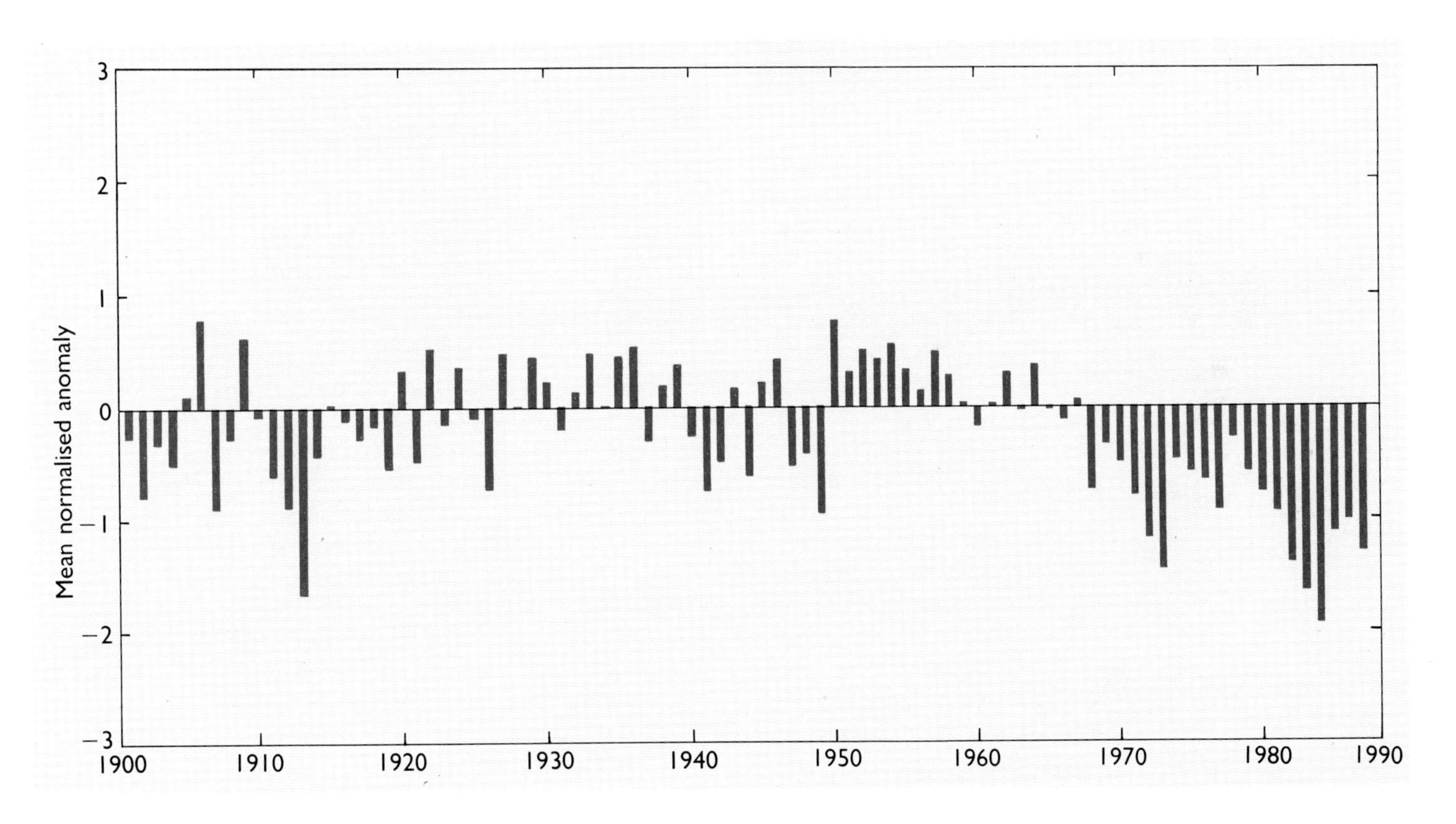

'For more than two decades, farmers in south-central Niger have lamented to development workers that "kasur mu, ta gaji", "the land is tired". Peasants in western parts of the country strike a more ominous chord with "laabu, y bu", "the land is dead".'

Sandra Postel, Worldwatch Institute

In Mauritania, a largely desert country where the Sahara meets the sea, the capital, Nouakchott, used to have about five days of dust storms each year during the 1960s. In the 1980s, Nouakchott suffered from dust storms on 80 days each year. In the 1960s, the city was home for 20,000 people – today, it is the base for 350,000, half of them officially classified as refugees. Dust from the region spreads far afield. 'Red rain', laced with dust from the Sahara used to fall only rarely over northern Europe. Up to 1980, there were seven occasions during the present century when this red dust reached England. In the 1980s alone, there have been ten downpours of red rain – a reminder of the suffering that continues in northern Africa. And the dust is also found in air masses crossing the Atlantic Ocean driven by the trade winds; it now reaches everywhere from northeast Brazil to Miami.

As harvests failed, the people who had been forced to give way to farming suffered. The nomads who had lived in balance with the environment were displaced and became refugees. Eighty five percent of Mauritania's people now live in cities, or in huddled squatter camps surrounding the urban centres. The farmers, trying to take from the environment what it could not give, were for too long encouraged to compound the mistake of trying to impose western-style agriculture in a semi-arid region.

The message is clear – brute technological force cannot buck the climatic trend. We need to be more subtle in our interaction with the environment. And we can learn a lot from traditional methods that make the most of what resources are available.

As Susan George warns in her book *Ill Fares the Land* (Writer and Readers, 1984): 'The slogan "caveat emptor" – "let the buyer beware" – has never been truer that for the case of Third World purchases of Western technology. As the Indian scholar A.K.N. Reddy has put it perfectly, technology is a carrier of the genetic code of the society that produced it. Once given, genetic codes are invariable. Those who purchase Western technology had best be prepared to adapt to it, because Western technology is not going to adapt to them.'

When the drought intensified in the 1980s – it didn't 'return' because it had never gone away – the consequences were inevitable, a tragedy of colossal dimensions. Lloyd Timberlake writes, in *Africa in Crisis*: 'In mid-1985, famine ... swept across Africa, from the Atlantic to the Horn and south to Mozambique and the South African bantustans. An entire continent was on the brink of collapse: the biggest disaster to strike the planet since World War Two devastated Europe. It was said that at least 30 million Africans in 20 nations did not have enough food to live on; that ten million had abandoned their homes and farms in search of food and water; many had abandoned their countries. But these figures are mere guesses. No one knows the numbers of dead and dying, least of all the governments responsible for their welfare.'

In spite of disaster relief – the efforts of charities such as Band Aid and Comic Relief – and the steady work of the more enlightened development agencies, the region may well have been better off if it had never received 'assistance' from the outside world. With fewer mouths to feed and a way of life more appropriate to the environment, famine may have ensued; but not on the same scale. All too often, inappropriate development schemes have made matters worse, rather than better. Lloyd Timberlake again: 'Fifteen years ago, the Sahel ... was ravaged by a similar tragedy, and its causes were largely identified by international experts. But the same mistakes have continued almost without change: that catastrophe taught the world little. Now, radical changes must be made, or "disaster" will become a permanent way of life for many Africans.'

Our responsibility may not end there. The conventional view is that the Sahelian drought is a natural phenomenon, one of the vagaries of climate that have occurred many times in the past. But what if it is not? What if the decline in subtropical rainfall described in Chapter 2 is the first serious evidence of the greenhouse effect at work? The possibility cannot be ruled out and, if true, it places the full burden of responsibility

for the Sahelian disaster firmly at the door of the industrialized world. It is the North that has produced the bulk of the greenhouse gases and it is the ultimate – and rather obscene – irony that it is the North that is, at least in part, benefiting from the shift in rainfall patterns of recent decades.

ADAPT OR DIE

We can learn a lot from the past. And perhaps the most striking and saddest lesson is that many of the problems we face today are of our own making. As global warming develops, we will have to learn to live in a radically different world. How will we face up to that challenge? Some might regard the prospects of higher temperatures as quite attractive – lower heating bills in northern climes, expansion of growing regions polewards, and so on – but this superficial assessment misses the point. For many inhabitants of our planet, climatic change and sea level rise will represent absolute catastrophe.

The crucial factor is that the rate of change will simply be too fast to allow adaptation to occur. Consider the fate of the natural world. Plants and animals live in all latitudes of the world, from the heat of the tropics to the cold of polar regions. It is easy to imagine that when the world warms there will still be a home for almost all forms of life. While that home may be a little further from the equator, that really isn't too

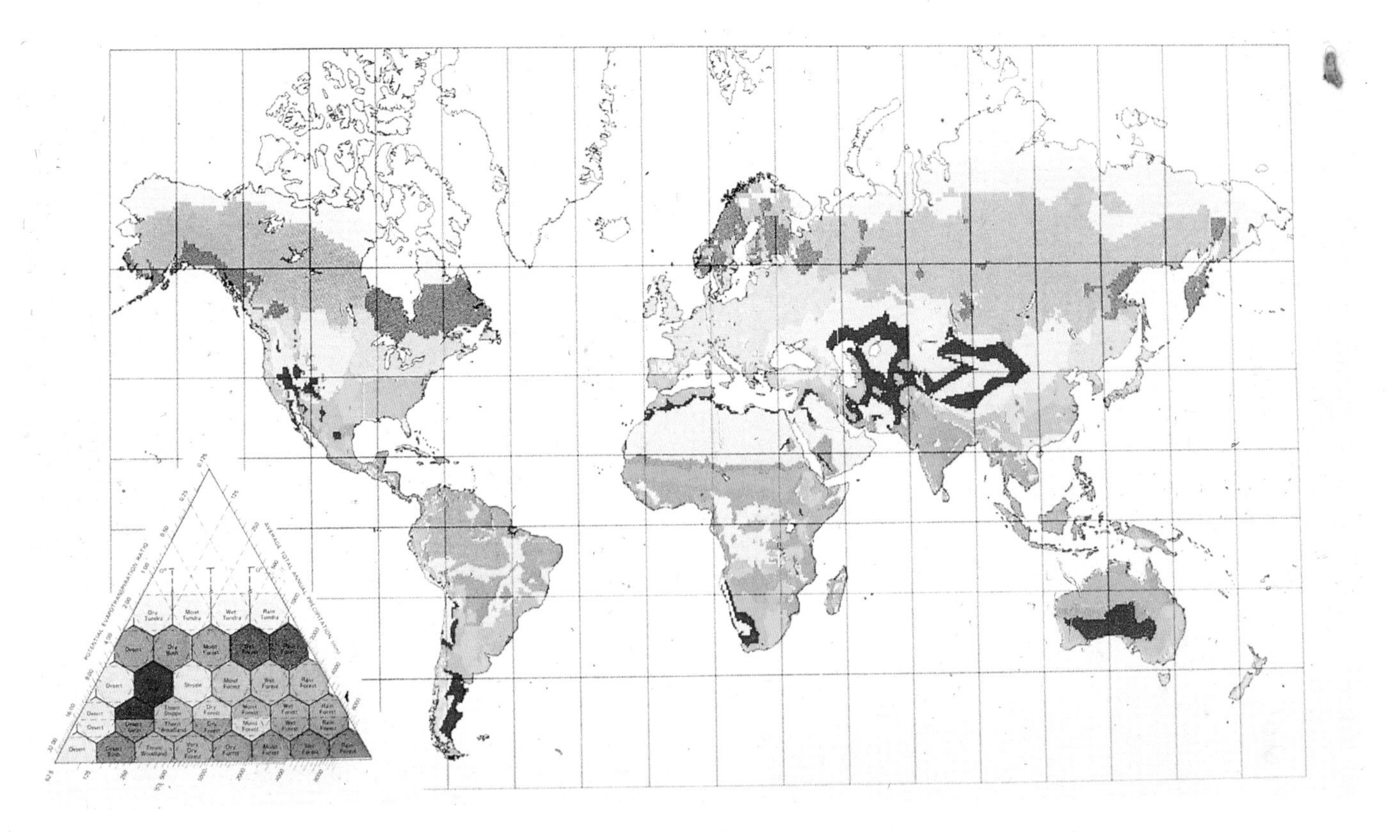

Key: The habitats referred to in the text are: red – *temperate desert bush;* cream – *cool temperate steppe;* light blue – *cool temperate forest;* purple – *boreal forest;* grey – *tundra*

The present day distribution of habitats based on regional variations in climate
Source: W R Emanuel and others

much of a problem – except, perhaps, for polar bears. After all, life on Earth survived the warming at the end of the most recent ice age. Some species were lost but many adapted successfully. There will surely be suitable environments for tigers, tortoises, oak and larch in a hundred years' time. The problem is that these habitats could be in quite different places from where the plants and animals live now. How will they get there?

Ecosystems can adapt and respond to gradual changes in climate, but as the speed of change increases it becomes much harder for them to keep up. The change in climate at the end of the latest ice age took thousands of years and carried the world into the state known as the interglacial. We are still living in this interglacial. But now, global warming threatens a comparable rise in temperature over mere decades. A temperature rise of several °C will cause a poleward shift of vegetation zones of several hundred kilometres. Animals and birds can respond relatively quickly to such a change – many bird species migrate over much greater distances as the seasons change each year. But forests and other vegetation cannot, and it is no good a bird heading off for more suitable climes if the trees for nesting, the fruit and seeds to eat, have yet to arrive. Even light seeds that are blown by the wind do not travel very far from their parent plants.

The Engelmann spruce provides a typical example of how slowly forests migrate. If the wind blew steadily from the south, a spruce forest could advance northwards at a maximum rate of only 20 kilometres a century. The seeds produced by each generation of trees must blow north, find fertile soil and grow into mature trees, whose own seed can then continue the expansion polewards. If climate zones were to move at a rate greater than 20 kilometres a century, which is certain if no action is taken to curb the greenhouse effect, then the forests will die out. And an ecosystem can only move as fast as its slowest component. Plants, animals and birds that depend on these trees will also be threatened with extinction. It is the *speed* of the climatic change, not just its size, that poses the threat.

'Biological diversity, already being reduced by various human activities, may be one of the chief casualties of global warming. Massive destruction of forests, wetlands, and even the polar tundra could irrevocably destroy complex ecosystems that have existed for millen-

nia. Indeed, various biological reserves created in the past decade to protect species diversity could become virtual death traps as wildlife attempt to survive in conditions to which they are poorly suited. Accelerated species extinction is an inevitable consequence of a rapid warming.'
Lester Brown, Christopher Flavin and Sandra Postel, *State of the World 1989*

Even in areas where the change is slow enough to allow some species to migrate, there will be problems. Migration paths may be blocked by natural barriers, like lakes and mountains, or artificial barriers in the form of cities and agricultural land. Human activity is already eroding threatened ecosystems through deforestation, urbanization and agriculture. Species weakened by pollution – for example, by agrochemical nitrates in rivers and acid rain – will be even less resistant to climatic change.

According to a survey conducted for the United Nations Environment Program, a substantial decline in forested area can be foreseen as early as the year 2000. As the deciduous forests of middle latitudes spread polewards taking advantage of the warmer climate, the coniferous forests of Canada and northern Europe, which thrive where there are snowy winters and short summers, will be severely eroded. This is not mere supposition – geological evidence shows that exactly this pattern of ecological change occurred over 120,000 years ago before the most recent ice age, when temperatures were just 2°C higher than at present. The 2°C warming mark could be passed within 30 years. At their equatorward margin, the temperate forests will be lost as increasing temperature and drought favour grassland – the prairies of North America and the steppe of Eurasia – rather than deciduous trees.

The oceans provide plentiful supplies of food for many peoples of the world. How will maritime ecosystems respond to global warming? Changes in ocean temperature, in ice cover, in the circulation of ocean currents will have a profound effect. Fish populations are likely to move towards the poles, although the pattern of change will be complex and unpredictable. In the long-term, the oceans may produce more food once the world settles into a new, stable hothouse regime. In the short-term, the change will prove adverse, disrupting the balance that maintains present-day food chains and ecosystems. Yields will be reduced. Fisheries will fail. The upheaval of the ocean climate associated with the record El Niño event of 1982 and 1983 resulted in losses of $270 million as the fisheries off the west coast of North America failed.

WHEN THE HARVESTS FAIL

Global warming will have a profound effect on food production. Fertile agricultural land in low-lying areas will be lost as sea level rises. Climatic change

CORNWALL COLLEGE LIBRARY

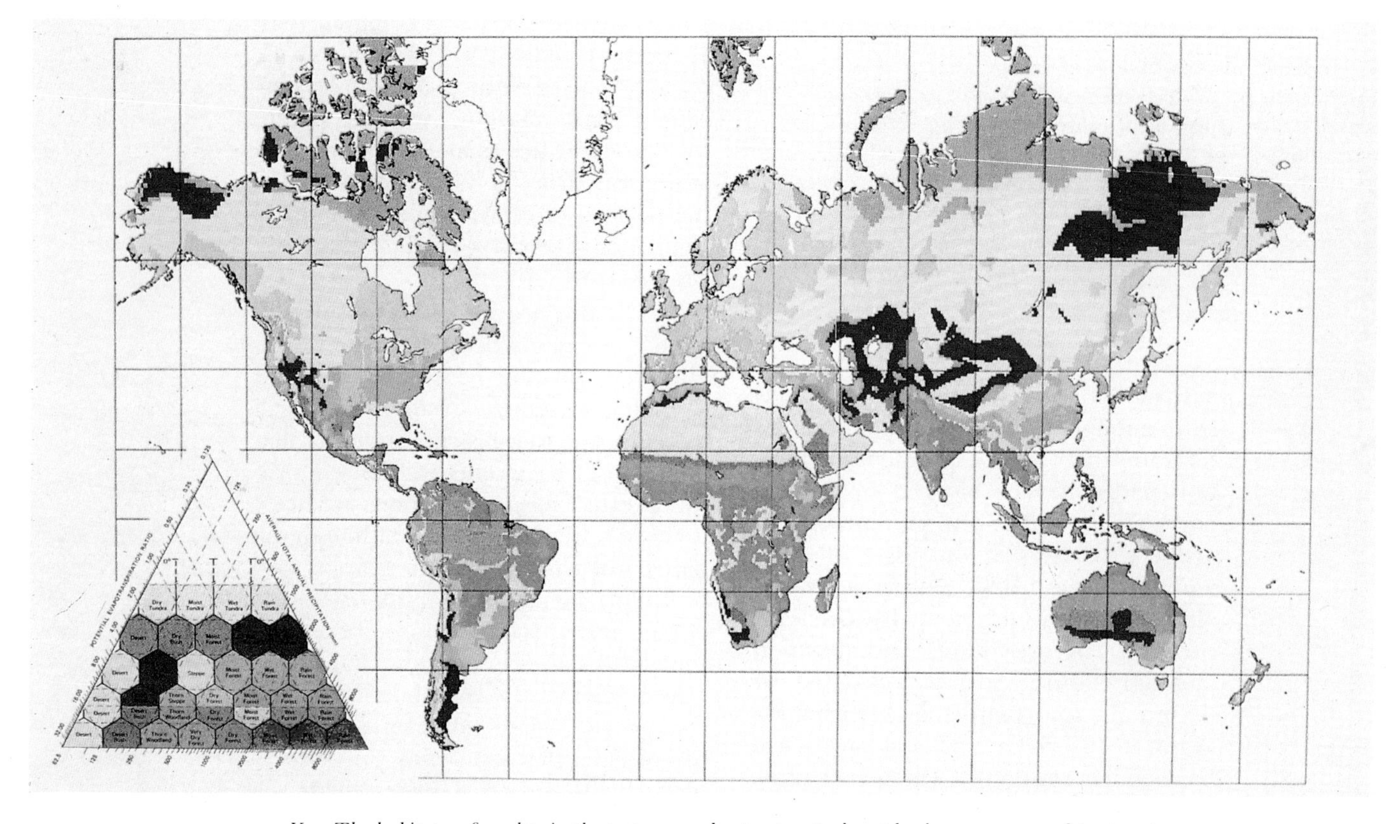

Key: The habitats referred to in the text are: red – *temperate desert bush;* cream – *cool temperate steppe;* light blue – *cool temperate forest;* purple – *boreal forest;* grey – *tundra*

Projection of the distribution of habitats by the early twenty-second century
Source: W R Emanuel and others

will reduce yields, the areas where crops can be grown will shift, nutrient levels in the soil will alter and farming practices will have to be revised. Before looking to the future, though, we should return to the present day and consider the current availability of food. It is the processes which lead to starvation today which will determine the ultimate impact of global warming on food security.

The world actually produces more food today than ever before – not just more food, but more food per head of population, even though population has been growing rapidly in recent decades. Between 1950 and 1985, cereal production grew on average at a rate of 2.7 percent a year, faster than population growth, and increased from 700 million tons a year to more than 1800 million tons a year. In spite of this spectacular success, at least 730 million did not have enough to eat in 1985 – 'enough', according to United Nations figures, to enable them to lead fully productive working lives. Food was available, but it was in the wrong place. For example, Africa and Asia do not produce enough cereals and root crops to sustain their present populations, yet over the world as a whole sufficient staples are produced to feed an additional one and a half billion people.

'The affluent living standards of people in rich countries are [therefore] directly connected to the poverty of most people in the world: we could not be so affluent if we were not taking most of the resources and gearing much of the Third World's productive capacity to our own purposes.'

Ted Trainer, *Developed to Death*

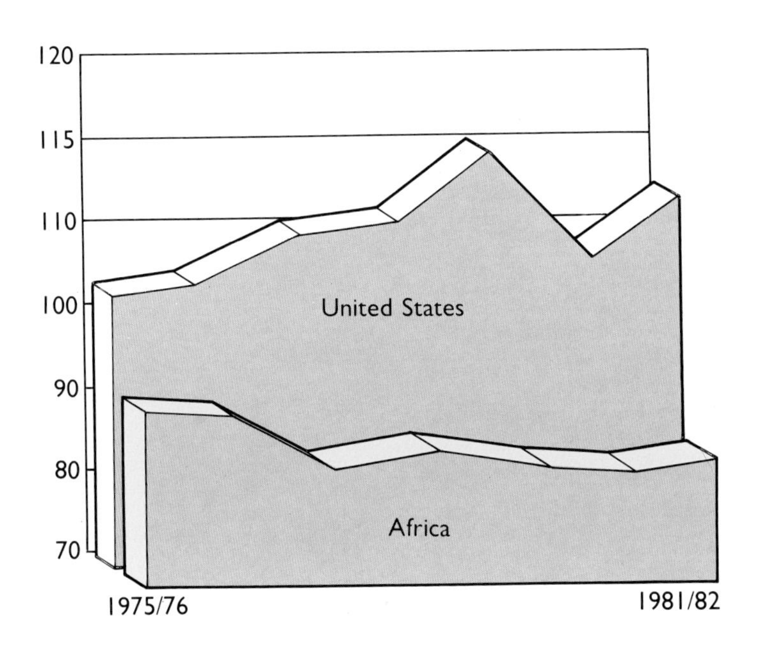

Trends in food production per head of population for the United States and for Africa Source: US Department of Agriculture/The Hunger Project

Why are people starving when we could feed the world?

One reason is that fashions in food have changed. More people have come to regard eating meat as a desirable, or even essential, part of their daily diet. Two-thirds of the increase in grain production in North America and Europe between 1950 and 1985 has gone to feeding animals. In the US, the average citizen now consumes over 100 kilograms of meat each year. In the UK, the corresponding figure is 75 kilograms. In India, it is just over one kilogram. If the grain that is fed to livestock were instead consumed by people, it would keep far more people adequately fed. One calorie of grain-fed steak costs at least ten calories to produce. Eating meat from animals that have been fed on grain is an extremely inefficient way of getting the nutritional value of the grain into your system.

Another reason is that the benefits of the 'green revolution' – the use of hybrid crops, agrochemicals and other technology to boost production – have been mixed. As the summary compiled by the World Commission on Environment and Development shows, some regions have benefited, others have not. In many cases, the failure to improve food production using new technology has occurred because it has been applied in unsuitable areas or inappropriate ways. All too often, western-style agriculture has been used as the model for develop-

The changing pattern of food production. Some regions have benefited from green revolution technology and other agricultural developments, others have not.

Africa
- a drop in per capita food output of about 1 per cent a year since the beginning of the 1970s
- a focus on cash crops and a growing dependence on imported food, fostered by pricing policies and foreign exchange compulsions
- major gaps in infrastructure for research, extension, input supply, and marketing
- degradation of the agricultural resource base due to desertification, droughts, and other processes
- large untapped potential of arable land, irrigation, and fertilizer use

West Asia and North Africa
- improvements in productivity due to better irrigation, the cultivation of high-yielding varieties, and higher fertilizer use
- limited arable land and considerable amounts of desert, making food self-sufficiency a challenge
- a need for controlled irrigation to cope with dry conditions

South and East Asia
- increased production and productivity, with some countries registering grain surpluses
- rapid growth in fertilizer use in some countries and extensive development of irrigation
- government commitments to be self-reliant in food, leading to national research centres, development of high-yielding seeds, and the fostering of location-specific technologies
- little unused land, and extensive, unabated deforestation
- growing numbers of rural landless

Latin America
- declining food imports since 1980, as food production kept pace with population growth over the last decade
- government support in the form of research centres to develop high-yielding seeds and other technologies
- inequitable distribution of land
- deforestation and degradation of the agricultural resource base, fueled partly by foreign trade and debt crisis
- a huge land resource and high productivity potential, though most of the potentially arable land is in the remote, lightly populated Amazon Basin, where perhaps only 20 per cent of the land is suitable for sustainable agriculture

North America and Western Europe
- North America the world's leading source of surplus foodgrain, though the rate of increase in output per hectare and in total productivity slowed in the 1970s
- subsidies for production that are ecologically and economically expensive
- depressing effect of surpluses on world markets and consequent impact on developing countries
- a resource base increasingly degraded through erosion, acidification, and water contamination
- in North America, some scope for future agricultural expansion in frontier areas that can be intensively farmed only at high cost

Eastern Europe and the Soviet Union
- food deficits met through imports, with the Soviet Union being the world's largest grain importer
- increased government investment in agriculture accompanied by eased farm distribution and organization to meet desires for food self-reliance, leading to production increases in meat and root crops
- pressures on agricultural resources through soil erosion, acidification, salinization, alkalization, and water contamination

Source: World Commission on Environment and Development

ment in environments or in social situations which cannot support growth of this kind. There is little point in relying on a tractor if you cannot get spare parts. New high-yield crops often require large amounts of agrochemicals which poor farmers cannot afford. And when western agricultural methods fail, the immediate response is to use more and more agrochemicals to maintain yields instead of dealing with the root

cause of the problem. 'It is analogous to taking stones from the foundation to repair the roof', comments Professor Garrison Wilkes of the University of Massachusetts.

But the main reason why people go hungry is simply that the food that is available is not distributed evenly. There is enough food produced each year to feed all five billion people on the planet adequately. But many of us eat

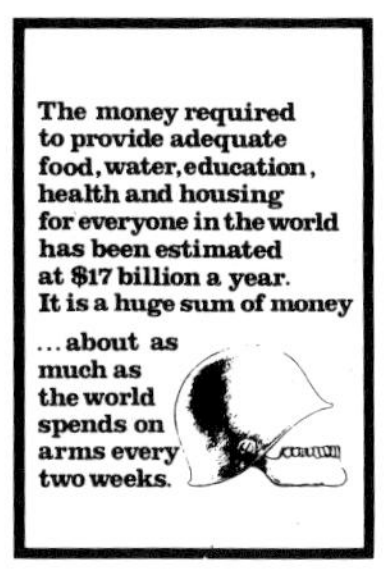

Source: Campaign against Arms Trade

far more than we need, while others starve. This situation is a product of many factors. But most come down to the lack of equity – justice and equality – in the world today. Food accumulates in the warehouses of the European Community, rotting until it is thrown away. Speculation on the commodities market leads to hoarding and forces prices up. Third World governments, trapped by debt, are forced to grow crops for export, cash crops which will earn foreign currency, neglecting the needs of their own people whilst pandering to the desires of the affluent North for exotic foods all year round.

As global warming intensifies, the impact will undoubtedly be experienced most acutely by the poorest people in the poorest nations. Agricultural production will improve in some areas, fail in others. But, wherever it occurs, climate stress will seek out the poor as the social and economic forces which are responsible for starvation in the present day shape the distribution of what food is available.

A variety of physical factors will determine how much food is available in decades to come. On the plus side, greater amounts of carbon dioxide in the air could boost the yield of some crops. Carbon dioxide is essential for growth, as we have seen in Chapter 1, and higher levels of carbon dioxide also help plants use water more efficiently. More carbon dioxide should mean higher levels of productivity. But some plants respond better than others. A particular crop may not respond as well as the weeds that are its competitiors, thereby reducing any gain. Moreover, laboratory experiments have shown that, although plants grow larger when additional carbon dioxide is present, there is less protein in the plant tissue, so more of the plant would have to be eaten for the same nutritional value. In any event, the direct effect of the additional carbon dioxide in the atmosphere is not likely to be strong enough to overcome its impact as a greenhouse gas, an agent of climatic change.

In the heartlands of North America and Europe, a temperature rise of 1°C alone would bring a decline in wheat yields of up to ten percent – and that is without considering effects on soil moisture availability. As we saw in Chapter 2, the drying out of the northern continental interiors of North America and Eurasia poses one of the greatest threats to global food availability. The USSR will also face reduced yields in the arid southern lands of Central Asia where water supply shortages already restrict agricultural production. The USSR could, however, ultimately benefit from the change in climate. Its vast latitudinal extent means that it could conceivably shift production from south to north in a way not open to the US and many European nations. But this will only be possible if fertile soils can be found, finance is available, and farmers can be persuaded to migrate.

There are other countries in higher northern latitudes that may be able to

Giant tomatoes grown in high levels of carbon dioxide.

take advantage of the change in climate and improve agricultural production. Canadian yields of winter wheat may increase significantly. In Scotland and parts of Scandinavia, crops such as oats may be grown at higher altitudes than is possible today, making more use of available land. For a few centuries in the Middle Ages when temperatures were around 1°C higher than in the middle of the twentieth century cereals flourished in Iceland and Norway. But these gains are unlikely to outweigh losses further south.

Rainfall will be the key factor, and current model predictions suggest that dry regions will become drier, wet regions wetter. Areas with a so-called 'Mediterranean' climate – Chile, Argentina, southern Africa, parts of Australia, much of California and the Mediterranean itself – could be hard hit by drought as global warming progresses. This will have a severe effect on production of crops such as wheat, maize and barley, and will enhance rates of desertification in and around areas already prone to this problem.

'We've treated the white man like a season. And we've tried to understand this new season that's upon us and now we're finally understanding that this season is almost like a pack of locusts on our lands. It's just eating everything and then it's going to move on. We're standing up now. We understand that we are the ones that can voice the voice of the land. We can voice our concern that your style of use is killing the land. We need to teach you, to humanize you, so that you can understand you just can't do this to the land.'

Gary Potts of the Teme Aushnabai Indians

Siberia, from woodland to wheatfield?

The greatest uncertainty over food supply projections affects just those regions where rapid population growth means that increased food production must be ensured. The changes in monsoon rainfall mentioned in Chapter 2 are bound to have a substantial impact on the Indian subcontinent, but nobody can say as yet whether this will enable more or less food to be grown. To the north, it does seem likely that higher temperatures will favour the production of rice near the northern limit of this crop, in Japan – but this is unlikely to help those starving in India or Africa. In the wetter regions, where higher rainfall is expected, yields may increase. But this will be to no avail if torrential rains and flooding wash away the crops. In regions such as the African Sahel, which are likely to become drier in the global greenhouse, food production can only suffer.

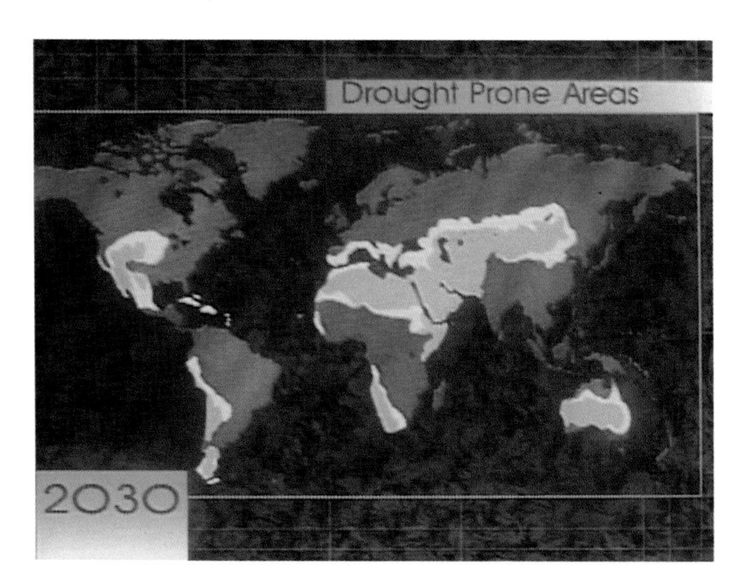

One thing in particular is quite clear in all this. The industrialized nations of the world are better equipped to weather this change in climate than the poorer nations of the Third World. Northern farmers are protected by grants and subsidies, by compensation in the event of harvest failure. They have the financial resources to invest in changing their farming systems, their crops and practices, and the technology to do so is readily available. The poorer farmers of the world, already hard-pressed to produce enough to support their families, will have no recourse when climate stress strikes, with no money to invest and their governments crippled by debt. Unless action is taken to reduce the disparity between rich and poor, there is a real risk that these climatic and agricultural upheavals will lead to international disputes, even war.

AS TENSION EXPLODES

In the words of the participants at the 1988 Toronto Conference, 'the best predictions available indicate potentially severe economic and social dislocation for present and future generations,

which will worsen international tensions and increase the risk of conflicts among and within nations.' How might climate stress be translated into open warfare? Once again, we are not making predictions – we are simply describing a few *possible* outcomes to draw out the enormity of the threat posed by global warming.

Migration is one of the most obvious sources of conflict. Imagine the situation in Bangladesh after flooding has caused a series of disasters, mass fatalities, and no-one wants to return to the fertile coastal land that was their home. Where will they go? Millions of people in search of survival. Despite the best efforts of the Bangladeshi government to deal with the crisis and aid from the international community, only force will prevent the desperate from journeying across frontiers into neighbouring countries where conditions, they believe, must be better. How will the government of India react when their own problems are compounded by hundreds of thousands of refugees?

> 'A one-metre rise in ocean levels worldwide ... may result in the creation of 50 million environmental refugees from various countries – more than triple the number in all recognized refugee categories today.'
>
> Jodi Jacobson, Worldwatch Institute

The new 'boat people' of the Mediterranean, fleeing the drought-ridden lands of northern Africa, will land in southern Europe. What welcome will they receive? Perhaps the early arrivals will be treated charitably; but as the trickle becomes a torrent it is likely that charity will be exhausted. This process can be seen at work in the response to the boat people of Southeast Asia, now passed from one country to another. Even with the best will in the world, there is only so much assistance that can be given. What will happen as illegal migration increases across the southern border of the US from Mexico? That border is already well on the way to becoming a militarized zone.

Mostafa Tolba, Executive Director of the United Nations Environment Program, gives a graphic example of how conflict over depleted resources may lead to tension, even warfare, between nations. 'I come from Egypt and Egypt is in a dry part of the world. The Nile gets its water resources from the uphills in Africa. We were on the verge of a collapse this last August when the rain fell. In fact, the floods that created all the problems in Sudan rescued Egypt from a disaster because we would have had to close down the turbines of the high dam which is a major source of our electricity.' What would happen if rainfall is reduced throughout the region and the flow of the Nile drops? Describing one possible response, he continues, 'If some country upstream puts a dam on the Blue Nile which brings almost 8

The Aswan High Dam on the river Nile.

percent of the water of the Nile down to Egypt, do you believe that 52 million people are going to sit back and wait, seeing themselves dying because of no water?' Butros Ghali, Foreign Minister of Egypt, considers that, regardless of global warming, 'The next war in our region will be over the waters of the Nile, not politics.'

Economic warfare may break out as old patterns of trade die and new ones emerge, as nations attempt to stockpile remaining resources while other nations without such resources pressure to make them available. Trade embargoes, sanctions and punitive import levies are increasingly employed to buttress foreign policy and these weapons will inevitably be used if national well-being is threatened by the impact of global warming. And when these tactics fail?

Bill Arkin, Director of the Institute of Policy Studies, sees the 'militarization' of the response to global warming as all too likely: 'I think our society is so infected with military thinking. You can see it already with drugs or AIDS. How do you deal with drugs? You interdict them. You close the borders. How do you deal with AIDS? There's a kind of fascism growing. You insulate yourself from AIDS. You take away the people who have AIDS – you segregate them. This is very military. It's not humane. It's progressive. If we start to deal with the deteriorating environment in the same way, insulating North from South in some way, looking at resources in the Third World as if they are potentials for our vitality? I don't have faith that governments in the North are going to say our major interest is going to be in improving the environments so that people elsewhere can live better. I think that governments in the North are going to look at their own national interests. There may be alliances that may be formed in that regard, but they are going to be alliances of those who live in the developed world, who want access to raw materials and want access to resources in order to continue to live in the style to which they have come accustomed. You will see the military establishment beginning to look at the Third World from a military standpoint.'

We have painted a depressing picture of the world's prospects over coming decades. But we have also pointed out that the causes of this impending crisis are readily identifiable – in the waste, inefficiency, overconsumption, injustice and misplaced priorities that are typical of our society. We will now present a more optimistic view of the future. One in which society acts to limit the threat of global warming – by adopting more sensitive methods of adapting to environmental change and by strenuous efforts to reduce emissions of the major greenhouse gases. These twin themes, adaptation and control, are the basis for any attempt to prevent the possible futures that we have sketched here from coming true.

International water disputes during the mid–1980s

Body of water	Countries involved in dispute	Subject of dispute
Nile	Egypt, Ethiopia, Sudan	Level of water flow
Euphrates, Tigris	Iraq, Syria, Turkey	Dams, water flow
Jordan, Litani, Yarmuk	Israel, Lebanon, Jordan, Syria	Water flow
Indus, Sutlei	India, Pakistan	Irrigation
Ganges	Bangladesh, India	Siltation, flooding
Mekong	Kampuchea, Laos, Thailand, Vietnam	Water flow
Paraná	Argentina, Brazil	Dam, land inundation
Lauca	Bolivia, Chile	Dam, salinization
Rio Grande, Colorado	Mexico, United States	Salinization, water flow, agrochemical pollution
Great Lakes	Canada, United States	Water diversion
Rhine	France, Netherlands, Switzerland, West Germany	Industrial pollution
Elbe	Czechoslovakia, East Germany, West Germany	Industrial pollution
Szamos	Hungary, Romania	Industrial pollution

Source: Worldwatch Institute

4

OUT OF THE HEAT TRAP

Gro Harlem Brundtland reflected the mood of many in June 1988 when she said: 'For too long we have thought of the atmosphere as a limitless good. We have been burning fuel and emitting pollutants, pressing aerosol buttons and blowing foam to our heart's content . . . Time has come to develop an action plan for protecting the atmosphere . . . Time has come to start the process of change . . . For too long have we neglected that we have been playing lethal games with vital life-support systems.'

Some politicians have countered calls for action to limit global warming with the claim that it would be best to hold off for a while until we understand the problem better. Unfortunately, the buildup of greenhouse gases is now proceeding so rapidly, and the rate at which these gases are being pumped into the atmosphere is itself increasing so rapidly, that there is no time left to

Gro Harlem Brundtland

'wait and see'. While further research will clarify what needs to be done, by the time understanding has improved significantly – many years according to most estimates – it will be too late to avoid significant impacts. Our children and their children and their children will be irrevocably committed to living in a drastically altered world. Indeed, some degree of further warming cannot be avoided as the climate system slowly responds to past emissions.

As Sherwood Rowland, expert on ozone depletion at the University of California at Irvine, has observed: 'What's the use of having developed a science well enough to make predictions, if in the end all we're willing to do is stand around and wait for them to come true?'. Steve Schneider of the National Center for Atmospheric Research reckons we should consider reducing the threat of global warming a form of planetary insurance. 'Why do we have insurance?', he asks those who claim the predictions must be more definite before action is taken. 'I don't know whether I'm going to break my leg, have pneumonia or not get sick at all – so I'll simply just wait. I won't spend the insurance premium until after I get sick. The problem is that then they won't insure you!'

The time to act is now. Policy measures take time both to conceive and to put into practice. Power stations have a lifetime measured in decades, not years, and decisions made today on which kind of fuel to use in new power stations will be affecting the environment well into the twenty-first century. And, as we shall see in Chapter 5, reaching the level of international understanding that will be necessary if substantial reductions in emissions are to occur is likely to take many years.

We can look at this issue from a different angle. All of the steps that must be taken to slow the buildup of greenhouse gases are desirable for other reasons. They amount, in effect, to better husbandry and more equal distribution of resources between nations and individuals; they amount to protective, rather than destructive, control over the global environment. So many of the means of limiting global warming make good sense anyway, regardless of the threat posed by the greenhouse effect. Reducing energy consumption would have direct economic benefits and would conserve dwindling fossil fuel reserves. Controlling vehicle and power station emissions would limit lead pollution, photochemical smog and acid rain. Banning CFC use would protect the ozone layer. Halting deforestation would preserve an invaluable resource, protect the way of life of indigenous peoples and reduce environmental degradation. The crucial question, perhaps, is not whether or not we should act now but rather – why haven't we acted already?

ACTION AND ADAPTATION

Gus Speth, President of the Washington-based World Resources Institute, has put the problem in perspective for those who still have doubts about the scale of the problem. 'A little warming,' he says, 'is no longer the issue. A big warming is.' A study carried out for the WRI by Irving Mintzer produced several projections of the warming that might occur if different sets of policies were adopted. In the WRI Base Case scenario, no new policies are implemented to slow the rate at which greenhouse gases are emitted. No effort is made to reduce deforestation, and energy policies are not influenced by any environmental considerations. CFCs continue to be released into the atmosphere. The net result is a global warming commitment of a further 1°C by the year 2000 (on top of the 0.5°C rise over the past century) and of at least 3°C above present levels by 2030. The word 'commitment' is used because of the time still required for the world to adjust to the new concentration of greenhouse gases; before the world can adjust the heat trap will inevitably intensify.

'It is vital to understand what these changes could mean ... To find conditions like those projected for the next century, we must go back millions of years.... If the greenhouse effect turns out to be as great as predicted by today's climate models, and if current emission trends continue, our world will soon differ radically from anything in human experience.'

Irving Mintzer, World Resources Institute

Irving Mintzer

'Yet this,' says Mintzer, 'is by no means the worst possible outcome.' If coal burning is actively encouraged, tropical deforestation intensifies and the Third World countries that have not signed the Montreal Protocol take up the use of CFCs as avidly as the US did between 1940 and 1970, then the world will be committed to a warming of 2°C by the year 2000. By 2030, we will be committed to a rise in temperature of 5°C – and in such a situation, it is likely that greenhouse gases will be emitted at a greater rate than ever before. By contrast, the WRI study looked at what they term a Slow Buildup scenario. Improved energy efficiency, less reliance on coal, less deforestation and controls on CFCs could, if implemented immediately, slow down the pace of warming considerably. But even so, the world is still committed to a *further* warming of 1.5°C by the year 2030, and by 2075 temperatures will be heading towards 3°C above those of today. 'We no longer have the opportunity,' concludes Mintzer, 'to avoid a substantial greenhouse warming altogether.' Even an heroic global effort will not prevent a change in climate larger than anything that has been experienced in human history – larger, indeed, than anything for 120,000 years. But the effort must be made. Steve Schneider of the National

Center for Atmospheric Research in Boulder, Colorado, puts it in a nutshell: 'Two degrees warming is not nearly as bad as five; and having it happen in 150 years is not nearly as bad as having it happen in 50 years.'

'Many important economic and social decisions are being made today on long-term projects . . . irrigation and hydro-power, drought relief, agricultural land use, structural designs and coastal engineering projects and energy planning . . . all based on the assumption that past climatic data . . . are a reliable guide to the future. This is no longer a good assumption.'
Conference Statement, *The Assessment of the Role of Carbon Dioxide and of Other Greenhouse Gases in Climate Variations and Associated Impacts*, October 1985, Villach, Austria

So, when responding to the threat posed by global warming, we have to consider not just controlling emissions but also what we are going to do to adapt – to a new set of climatic conditions, to rising sea level. Change is inevitable. And if we are going to adapt without undue suffering, action must be taken soon. The timescales involved in the planning and construction of coastal defences, for example, are of the order of decades. The Dutch have only just completed the improvement of their sea defences triggered by the 1953 floods. Irrigation and hydroelectric power schemes are not the sort of projects that can be started up, let alone completed, in a year or two. It takes around ten years, judging from past experience, to change the emphasis of a national agri-

cultural system towards new crops and different kinds of land use. Breeding the new strains of grain and making them widely available also takes ten years. For some crop species, the time needed is longer. And there may not be enough technically-trained people around to do all these jobs. If the scientists have to be educated and trained first, that will add many more years to the programme.

The time for debate about whether the world will warm in the twenty-first century is over. That can now be considered a fact of life. Although the discussion continues concerning how great that warming will be, and how soon the first substantial impacts will occur, that debate is of secondary importance. The time for action to minimize the impact of the warming is already upon us. If we wait, it will be too late – moreover, none of the actions we need to take will do us not to be quite so severe, or quite so rapid, as the worst scenarios suggest. We will be better off for implementing these policies, whatever happens to the weather.

HELPING NATURE HELP HERSELF

The most effective way to help natural ecosystems to cope with global warming is to allow, or encourage, them to maintain and develop diversity, a broad variety of living things within each habitat. That is the way in which the Earth's living systems have coped in the past with climatic variations as extreme as a full ice age or the global warmth in which the dinosaurs thrived. Diversity means flexibility, keeping options open, and it increases the resistance or resilience of an ecosystem to climatic stress. Flexibility and resilience are the keys to coping with change, whether on the part of humanity or nature. And flexibility and resilience are present in abundance in ecosystems such as the rainforest. When the forest is cleared, what is left is a very simple ecosystem, far less able to withstand the vagaries of climate. By leaving the rainforest alone or restoring it to the extent that is possible, we would be helping it to adapt to the growing greenhouse effect – and by leaving the rainforest alone we would, simultaneously, be reducing the rate at which the global warming will occur.

To what extent society will intervene to protect ecosystems from the impact of environmental change is a moot point. It is extremely unlikely, for example, that any strenuous and expensive efforts will be made to preserve ecosystems by building defences to protect biologically-rich wetlands from the rise in sea level. But other cases are less clear cut. What happens to parks and re-

serves where rare and endangered species are protected today? Very often, these parks are surrounded by agricultural land, or even by urban development. When the climate changes so much that the region where the park is located can no longer support the species it is there to protect, who will pay to move the inhabitants of the reserve, lock, stock and barrel, to a new location? Or will we simply write off the rare species that we have already made a major effort to protect? It is considerations such as these which will determine how much we are prepared to meet the costs of control as opposed to the penalties of adaptation. Control, in that it largely involves altering our own actions rather than further tampering with nature, is likely to prove the easier option.

And the costs of losing ecosystems such as the rainforest will be high. It is important to understand why they are so valuable, not just for their scientific interest but to human society at large. Living resources – plants, animals and micro-organisms – are an indispensable part of our future. Improved yields of crops, for example, come from breeding selected strains to produce maximum yields under a particular set of conditions. But when the temperature or rainfall deviates from the expected 'normal', those highly bred crop strains are much less productive. The solution to this problem is to go back to nature, and select other varieties to breed from to achieve high yields under the new normal conditions. Or better, the way things are going now, to develop strains that will give fairly good yields under a wide range of conditions rather than huge yields in good years and nothing when conditions are slightly less than optimum.

Wild species contain a vast variety of genetic possibilities – you can see this by looking at the enormous variety of dogs around today, from chihuahua to great dane, all of which have been bred from a common, wolf-like ancestor. The genetic material of those ancestral wolves contained the potential for every dog alive now. The DNA of plant species in the tropical forests contains just as much potential, and makes contributions to agriculture, medicine (in the form of new drugs) and industry (in the form of new raw materials) worth billions of dollars every year. These huge stocks of biological diversity are under threat, just at the time when the science of genetic engineering is developing the potential, with due care and control, to make even more effective use of them.

All natural ecosystems are valuable, in the direct and selfish sense of being useful to us. But the rainforests that cover just six percent of the land surface of our planet contain at least 50 percent, and possibly 90 percent, of all species of life on Earth. If deforestation in the Amazon basin continues at its present rate to the end of this century, at least 15 percent of all plant and animal species will be lost. Any one of those plant species, if it had survived, might have provided the basis for new drugs or new foods.

The report, *Our Common Future*, puts it succinctly: 'Species that are important to human welfare are not just wild plants that are relatives of agricultural crops, or animals that are harvested. Species such as earthworms, bees, and termites may be far more important in terms of the role they play in a healthy and productive ecosystem. It would be grim irony indeed if just as new genetic engineering techniques begin to let us peer into life's diversity and use genes more effectively to better the human condition, we looked and found this treasure sadly depleted.'

At this time, the key factor in helping nature to help herself is still the need to bring these points home to policymakers, and to the people who can influence their decisions. As Anne and Paul Ehrlich dryly observe in their book *Earth*: 'The potential values to be found in the world of other organisms are essentially non-existent to conventional economists – and to businesspeople, politicians and others who share their myopic view of reality. They would recommend turning the Amazon Basin into a parking lot if the present value of the parking fees were greater than that computed for the possible cancer cures or life-saving new crops that would be lost forever by clearing the species-rich forests.' It is self-interest, not just a woolly desire to protect 'the environment', that makes it essential to minimize our interference with natural ecosystems. If we do, then not only will we be reducing the rate at which greenhouse gases build up, we will be enabling those ecosystems to cope with the changing climate more effectively, and we will be preserving the genetic reservoirs that we may need in order to help agriculture adapt to the rise in temperatures and the changes in rainfall. Myopic policymakers must be made to put on their spectacles and take a look around them.

'The potential values to be found in the world of other organisms are essentially non-existent to conventional economists – and to businesspeople, politicians and others who share their myopic view of reality. They would recommend turning the Amazon basin into a parking lot if the present value of the parking fees were greater than that computed for the possible cancer cures or life-saving new crops that would be lost forever by clearing the species-rich forests.'

Anne and Paul Ehrlich, *Earth*

HELPING AGRICULTURE, HELPING OURSELVES

At present, agricultural planning takes no account of the possibility of climatic change. Individual farmers' methods, national plans, and regional policies like those of the European Community are based on our experience of past history, and the expectation that the weather we are used to will be the weather of the future – that history will repeat itself. Yet one thing we now know for certain is that history will *not* repeat itself. And some allowance for climatic change can be made at all these levels, at little cost.

There are two key requirements in planning agriculture in the global greenhouse. First, simply to increase the resilience of agricultural systems, both by planting varieties of crops that will do reasonably well in a broad range of weather conditions, and by planting a mixture of crops so that even if some fail others will survive. Unfortunately, this kind of policy is the direct opposite of the shift that has occurred in recent decades, towards huge prairie fields planted with a single crop, a highly bred strain genetically tuned to produce maximum yields under optimum conditions but which cannot cope with change. Second, more subtly, planners should take on board the opportunity to make use of the fertilization effect of carbon dioxide by developing and planting varieties that will thrive as the concentration of carbon dioxide increases.

Some adaptation to the changing climate will occur in the normal course of affairs. As frosts become rarer, farmers will increasingly take a chance on planting earlier to allow crops to develop sooner during the spring months, and increasingly they will get away with the risk. As the benefits pile up in the warehouses and bank accounts, more farmers will seize the opportunity. The timing of irrigation for crops in the fields can also be altered relatively easily, provided water is still available. That, however, is a big proviso. The flow of the Colorado River, for example, will decrease by a quarter percent for a 1°C rise in temperature accompanied by a ten percent decline in rainfall, and the Colorado is already hard pressed to provide the water farmers want from it.

The Hoover Dam

Farmers must be prepared to use water more efficiently, and not allow it to run to waste. And, on only a slightly longer timescale, drainage schemes can be altered, and windbreaks constructed, as the message gets through that events like the drought of 1988 in the US are no longer isolated freaks but will increasingly become the new norm.

As the change in climate progresses, though, new crop varieties will *have* to be introduced, and patterns of land use *must* change. These changes will take a long time to get in to operation, and it is imperative that the first steps in these directions are taken now. A main priority must be to increase funding for research into the regional impacts for the global warming, so that we know in more detail what we should be planning for. Then, we can consider what the options are. There are some specific

Young seedlings grown in the laboratory

long-term projects where research begun now will repay itself many-fold in the twenty-first century. These include work on pest control and disease resistance. Pests and diseases that we think of now as tropical will spread further from the equator as the world warms, affecting plants and animals that have never had to cope with them. Breeding resistance into these crops in advance will help to mitigate the impact of these problems. And plants that can tolerate salt are already being bred. Developing this research further will yield strains that can survive even where salt water penetrates beneath agricultural land as sea level rises.

In all these cases, the industrialized world is, once again, better geared to adapt to the change. It is the poor farmers of the Third Word who are most at risk. The North will have the resources to breed new crops, restructure farming systems, subsidize farmers in times of trouble. Paradoxically, one of the best ways in which the poor of the Third World can help themselves is by abandoning the farming practices they have copied from the North – the very techniques that help insulate agriculture in the developed world against moderate climatic change – to concentrate on more traditional, and more appropriate, methods.

The poorest countries could benefit immediately from the re-introduction of traditional practices neglected following the adoption of western farming methods. This would cost little and could be implemented very quickly. Plant cover should be maintained throughout the year by sowing the same crop at different times and by using different varieties of crops. Bare soil, when and where it does occur, should be covered by mulches of leaves, brushwood or even plastic sheeting, to trap moisture and protect young plants. And all available sources of nutrient, from sewage to ash and compost, should be utilized. Such practices are commonsense, and seem almost too embarrassingly obvious to be worthy of active promotion. But all too often in the Third World, and especially as control over agriculture has moved out of the hands of small subsistence farmers and become a matter of agribusiness, producers have abandoned such good habits and tried to emulate the technique of the North, with larger fields and

single crops, and increasing use of mechanical aids and chemical additives. In the short term, this is more profitable; in the long term, it feeds fewer people. It has been proven time and again that this is not an effective way to feed the population of a poor country.

It also happens that the old-fashioned way of farming is more labour-intensive than modern, mechanized methods. This is sometimes cited as a point against it. But one of the increasing problems in the Third World, as population rises, is that of unemployment, with people drifting away from the land and into huge urban regions, where they have to be fed anyway and all too often still cannot find work. There is no lack of people who could be involved in commonsense agriculture. There is, admittedly, a lack of enthusiasm for that lifestyle, but it is surely better than life in some of the shanty slums that now surround Third World cities. Farming could be made attractive once again, but only by reversing the process of inappropriate urban-based development that has destroyed the structure of so many societies. It will not be easy, but anything which can be done to increase the amount of food available to the poor will provide some reserve against the vagaries of climatic change.

Food reserves on a larger scale will be an essential buffer as climate changes, as crops fail in some regions while they thrive in others. In the humid tropics, half of the food stored at the local level may be lost each year through the attacks of insects, rodents and fungal diseases. Even in semi-arid Africa, a quarter of the harvest is lost to non-human consumers. Simply providing better containers for the grain at farm or village level could halve these losses. The issue of food distribution must also be addressed, a problem highlighted by the difficulty of taking relief supplies to starving people in Ethiopia and Sudan in the 1980s. Many of the countries

where people are going hungry have no real roads, let alone other transport systems. When the harvests fail, it is easier for people to move to centres where they can be fed than for food to be taken out to individual families in their homes – even then, many die on the road to relief camps. This prolongs the crisis. The abandoned fields cannot recover and become productive again without the farmers there to tend them, while the farmers cannot leave the feeding centres or they will starve. It is crucial to find ways to take food to the people when crises like this occur.

The same problem exists on a larger scale – it is nothing less than an outrage that malnutrition and starvation still occur in Africa while surplus food supplies accumulate, and even rot, in Europe. There are, in fact, well-equipped organizations that have the transport to do the job of moving food where it is needed, both on a global and on a regional scale, and who could be made available as climatic crises intensify, if the political will were there. They are the armies, the air forces and the navies. Their skills might also be put to good use in another role, the defence of low-lying regions from invasion – by the sea.

HOLDING BACK THE SEA

The most obvious way in which we can help ourselves to adapt to the hothouse of the twenty-first century is by improving sea defences and coastal policies. Where sea defences already exist, they are designed to cope with extreme events – not the normal level of high tide on a quiet day, but the anticipated worst possible surge that washes inland when a storm lashes against the coast. As well as the rise in sea level which will occur over the next few decades, there will be an increase in storminess in many regions as the world warms. So, if a sea defence is considered adequate today, that does *not* mean that we can necessarily improve the barrier to cope with, say, a ten centimetre rise in sea level simply by making it ten centimetres higher. In most cases, greater protection will be required to guard against the risk of storm surges. And as sea level continues to rise, even more protection will be needed. Just what level of risk to plan for under these circumstances is a difficult decision.

The way in which coasts can be defended are obvious, but expensive. They include the construction of dikes and sea walls, and raising the level of the land near the coast with infill. Other options also have military analogies – a planned retreat from threatened regions will be better than the rout of emergency evacuation as the floods strike. But moving homes and industries inland, although less costly in the long run than leaving them to the mercy of the rising sea, is expensive in the short term. Once again, this poses a particular problem for the Third World, although it is also a factor in planning the defences of less populated regions of the developed world. 'First aid' may be all that can be applied to those regions, in the form of flood warning systems and contingency plans to cope with the floods that will occur increasingly often, and with increasing severity, as we move into the next century.

The threat of rising sea level should also be taken into account when planning new development projects – an existing proposal for a dam on the Ganges upstream from the Delta ought now to be reconsidered, for example, because the presence of the dam will hold back silt that would otherwise wash down into the delta and build up natural barriers against the sea. A combination of the dam and rising sea level will cause worse flooding than rising sea level alone. It is unfortunate that, to date, the multilateral development agencies such as the World Bank have paid little attention to the implications of global warming.

In spite of the best efforts that could conceivably be made to try to adapt to rising sea level and to hold back the floods, the pressure of increasing population and the lack of alternative agricultural land will mean that many millions of people will be at risk as sea level inches upwards. The most effective way to reduce the misery that will result is by reducing the rate at which sea level will rise. That means reducing the rate at which global temperatures will rise – by controlling the rate at which greenhouse gases are emitted. As

Florida, today (above) *and after a sea level rise of over 3 m* (below)

the WRI Slow Buildup Scenario we described at the beginning of the chapter demonstrates, this can be done.

MAKING THE MOST OF WHAT WE'VE GOT

Environmental considerations rarely play a significant part in decisions concerning future energy policy. Yet, in the words of the World Commission on Environment and Development, 'choosing an energy strategy inevitably means choosing an environmental strategy'. Ninety percent of all the commercially-generated energy used by humanity today comes from fossil fuel. Anything we can do to limit the use of energy will slow the increase in global temperatures caused by the greenhouse effect. That means, first and foremost, using energy more efficiently – making the most of every ton of coal and barrel of oil that is burned. By making our houses, factories and offices more efficient at trapping heat, through better insulation, we will lessen the increase in heat trapped by the whole planet under its greenhouse atmosphere. It has been suggested that we might remove carbon dioxide from burnt gases before they leave the power station chimneys but this is, unfortunately, no solution: it takes more energy to drive the chemical reactions that absorb carbon dioxide. More energy means burning more fuel, and you are back where you started.

At present, the growth in global energy consumption is running at a rate of just under two percent a year. This is less than half the rate at which energy use was growing in the early 1970s, before the dramatic rises in oil prices during that decade. The oil crises of the 1970s broke the historic link between economic growth and growth in energy demand, and showed just what is possible, in terms of energy saving, when nations are forced to find solutions. But even these solutions have barely scratched the surface of the savings that are possible. A recent World Resources Institute study, *Energy for a Sustainable World*, has shown that the growth in demand for energy could be reduced still further, even allowing for increasing population and a better standard of living for people in the Third World. It suggests that the amount of energy used by each person in the developed world could actually be halved by the year 2020, without any decline in living standards.

If you find that hard to believe, think again. These savings are possible because we have been so profligate in our use of energy in the past. In the UK, for example, less than 40 percent (not even half) of the energy released by burning coal, oil or gas is used by the consumer. It is hardly surprising that this rate could be improved upon. Although there are limits, set by the laws of physics, on how much of the energy from burning coal, say, can actually be converted into usable electricity, the waste heat from power stations, dissipated into the air in huge cooling towers to-

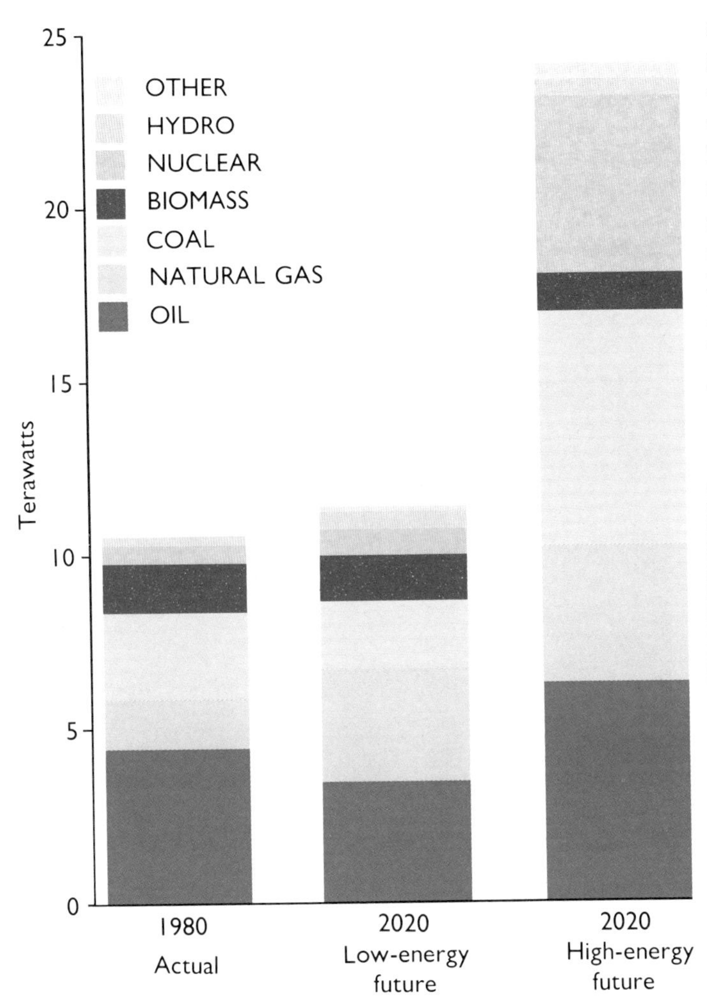

Alternative projections of global primary energy use broken down into various energy sources
Source: World Resources Institute

day, could provide up to 30 percent of Britain's requirement for space heating and hot water, saving up to 30 million tons of coal a year, if it were diverted through piped systems into homes and factories. This cannot be done by converting existing large power stations, which are too far from urban centres; but it could be done if new power stations were to be built on a smaller scale, closer to centres of population, to meet these twin requirements of combined heat and power. Unfortunately, the British government spends £1 million a year on research into coal technology, £170 million a year on nuclear research. A clear case of misplaced priorities.

Housing and office buildings have seldom, in the past, been designed and built with efficient thermal insulation. When it gets cold in winter, we simply turn the central heating up. But super-insulated houses have already been developed in Sweden and in the US. They reduce the amount of heating required in winter by nearly 90 percent. As a bonus, they also keep the interior of the building cool in summer, reducing the need for air conditioning. Planting a few trees in strategic locations around a house can also reduce air-conditioning bills considerably, and the trees will also take carbon dioxide out of the air.

'The fastest, least expensive, and above all, most effective response to greenhouse warming induced by fossil fuel combustion is to curtail the emission of carbon dioxide by improving the energy efficiency of the global economy.'
Bill Keepin and Gregory Kats, Rocky Mountain Institute

Transportation accounts for close to a third of carbon emissions in North America; it is one of the major sources in the industrialized world. The number of cars on the world's roads has grown from 50 million in the late 1940s to close to 400 million during the 1980s, and continues to rise. The most direct way of

making transportation more efficient and reducing carbon dioxide emissions would be to promote the use of public transport, car-pooling, bicycling, or walking. But fuel efficiency can also be improved. Fuel consumption figures come down with every new generation of vehicles put on the roads today, and this progress could be maintained for many more years. One practical difficulty, though, is that vehicles are kept on the roads before being scrapped much longer in the Third World than in the North. This is a good thing in many ways, reducing the consumption of resources such as steel (and the energy used in turning those resources into vehicles). But it does mean that it takes longer for improved, fuel efficient vehicles to become part of the pool of Third World transport. And old, poorly maintained vehicles tend to use more fuel. This is an area where the issue of aid and assistance, discussed in the following chapter, could prove crucial.

The industrialized nations have the greatest potential to make savings of these kinds, because they use – and waste – most energy today. In 1980, global energy consumption was about 10,000 billion watts (ten terawatts). If the amount of energy used per person in each part of the globe stayed the same, rising population would increase this figure to fourteen terawatts by 2025. But nine terawatts would be generated in developed countries, and only five in the Third World. If everybody in 2025 used energy at the rates it was used in the industrialized world in 1980 demand would be 55 terawatts, more than five times the 1980 figure. That simply is not sustainable, because of the enormous impact generating so much energy (which could only be done through vast increases in the burning of fossil fuel) would have on climate. We have to avoid this particular future.

But Third World nations *must* use more energy each year to improve the living standards of their increasing populations. The WRI scenario envisages a 30 percent increase in the energy use per head in the Third World to the year 2020. So the best we can reasonably hope for is to reduce energy consumption in the industrialized world to compensate for the increasing demand in the Third World – but the relatively modest growth envisaged by the WRI scenario can only be kept down to such a level if the Third World itself adopts efficient means of energy generation. Again, this can only be achieved with active cooperation and assistance from the industrialized countries. Brazil provides a good example of the potential for savings, and the problems of achieving them. Studies have shown that an investment of $10 billion in more efficient refrigerators, vehicle motors and street and building lighting would not only cut the nation's consumption of fuel by tens of percent but would make sound economic sense, leading to a saving of $44 billion by the year 2000 as less fuel is burnt. But Brazil does not have $10 billion to spend on such a project. If it is to be carried through, it will require assistance from the richer north.

'In Brazil, for example, a recent World Bank study indicated that an investment of about $10 billion would replace an investment in energy supply capacity worth about $44 billion. Nonetheless, the Bank continues to invest about 99% of its energy sector loans in increasing energy supply rather than improving energy efficiency.'
Irving Mintzer, World Resources Institute

There are similar problems with less sophisticated systems of energy use. In the poorest parts of the world, open fires provide the main source of heat for cooking. A great deal of the heat from these fires is wasted. If each household had an efficient stove, the need for fuel would be cut in half. The cost of providing efficient fuelwood stoves to every rural household in the Third World would be about $1 billion a year. Poor countries do not have the finances for this. Compared with the threat of global warming, developed countries ought to regard this sum as a small price to pay in reducing the rate at which forests are destroyed. As a bonus, such a programme would help to remove what has been referred to as 'the tragic paradox of poverty'; in *Our Common Future*, the World Commission on Environment and Development highlights this paradox: 'The woman who cooks in an earthen pot over an open fire uses perhaps eight times more energy than an affluent neighbour with a gas stove and aluminium pans. The poor who light their homes with a wick dipped in a jar of kerosene get one-fiftieth of the illumination of a 100-watt bulb, but use just as much energy.'

In spite of these opportunities, there are still major problems in trying to stem the growth in the use of fossil fuels in the Third World. The biggest problem concerns China. There, a hastening race to industrialize and raise living standards is being fuelled, literally, by coal. China has huge reserves of coal, and the cheapest and quickest way to develop the nation's economy is to burn lots of that coal in power stations, steel mills, and so on. China is now the main coal producer in the world, extracting (and burning) 900 million tonnes of coal per year. 'This,' says Huang Hua, Minister for Environmental Affairs, 'is not because we like it; this is dictated by the nature of our energy reserves. In my opinion, China will not be able to change this energy structure in a fairly long period.' But China, like the UK and the US, can use the energy generated by that coal more efficiently, with the aid of combined heat and power plants, better insulation, and so on. The time for the industrialized nations to help China down the path of energy efficiency is now, with aid and the transfer of technology and expertise, before

Huang Hua

progress develops too much momentum down the wasteful route pioneered by Europe and North America.

Energy efficiency is well established as the cheapest 'source' of energy for the future. It can be ten times cheaper to save a unit of electricity than to generate a new unit. And saving this unit has little if any impact on the environment. After using our energy efficiently, the next best thing we can do to slow global warming is to alter the mix of fuels used – away from fossil fuels, if possible. Where this is not possible, some benefits can be gained by using different fossil fuels instead of coal.

CHANGING THE FUEL MIX

All fossil fuels release carbon dioxide when they burn, but gas and oil release less carbon dioxide, per unit of energy produced, than coal. In the short term, it would be possible to slow the warming of the planet by burning less coal and more oil and gas. But since the global reserves of coal are much bigger than the reserves of oil and gas, this could only be a short-term remedy, giving us a breathing space while other sources of energy which do not produce carbon dioxide at all are developed.
Oil represents just 16 percent of known fossil fuel reserves, and gas only about ten percent. All the rest – just about three-quarters of known reserves – is coal. At present rates of use, known reserves of coal would last for 3000

years. There is enough gas to keep us going for 200 years. But oil production, which is still growing today, is likely to level off early in the twenty-first century, and then decline. As supplies decline, the price will rise and consumption will ease back, so most forecasters see oil contributing a diminishing amount of our energy throughout the twenty-first century. Burning oil produces less carbon dioxide than burning coal, and burning natural gas produces less carbon dioxide than burning oil. However there is only limited scope for expanding the use of oil and gas at the expense of coal: two-thirds of the energy produced commercially in the world in the late 1980s already comes from oil and gas. Moreover, oil, in particular, is needed as a source of raw material in the chemical and plastics industry, and many people already argue that it is too valuable, and too scarce, a resource to be wasted by burning it at all. In order to get that breathing space, we will have to rely mainly on switching from coal to natural gas. According to the WRI Slow Buildup scenario, this could bring about a reduction in the amount of coal burnt each year of 20 per-cent by 2020 – provided that the use of other forms of energy is also increased. According to Lester Brown and his colleagues at the Worldwatch Institute, 'If policymakers do not grasp the link between energy efficiency, renewable energy, and global warming, climate stabilization will not be possible.'

A wind farm in California (above) and solar panels in Somalia (left)

Many of the non-fossil fuel forms of energy production – the 'renewables' – are already familiar. They were widely advocated during the oil crises of the 1970s, and as a result of those crises some research and development has been done to make them practicable on a large scale. Solar power, hydroelectricity, wind and wave power, and geothermal energy tapped by drilling deep into the Earth's crust are all systems which do not involve burning anything or releasing any carbon dioxide. The Worldwatch team again: 'Hydropower supplies about 21 percent of the electricity worldwide, displacing 539 million tons of carbon that would otherwise be emitted each year. Solar collectors are a major source of hot water in Israel, wind power has taken hold in California, and geothermal energy is a major electricity source in the Philippines.'

Methane, which we have already discussed as a greenhouse gas, is another 'alternative' source of energy, and burns easily (methane is, indeed, the main component of natural gas). As methane is produced by the action of bacteria on animal and vegetable wastes, methane-based systems can be particularly useful on farms, both in the developed world and in the villages of the Third World. They are often referred to as 'biogas' systems. Although burning methane does release carbon dioxide into the air, this is not a major problem because the carbon in the gas is part of the natural cycle of life and would have been released relatively quickly anyway. Carbon is taken out of the air by plants during photosynthesis, get eaten by animals, is excreted, and is then converted into methane. If the ecology is in balance, that methane will react with other gases in the air to make carbon dioxide once again, ready for the carbon to do another round through the cycle. Problems with methane as a greenhouse gas arise because human intervention has changed the natural balance, tipping it in such a way that more methane than ever before is now getting into the air. Since we are producing so much more methane, it is better to burn it than to burn coal and thereby add additional carbon dioxide to the air, as well as extra methane. In fact, since each molecule of methane is 20 times more efficient than a single molecule of carbon dioxide at trapping heat, burning methane, wherever and whenever it is found, directly reduces the greenhouse effect.

Similar arguments apply to the use of biomass – plant material – as fuel. Provided that the plants are harvested in a sustainable way, and new plants are grown to replace them, we are simply cycling a pool of carbon around the atmosphere and biosphere. Indeed, since the carbon gets in to the plants through photosynthesis, a process which depends on the action of sunlight, burning biomass is simply a way of using solar energy. The same is true of burning fossil fuels, of course, as these fuels are the remains of living things, deposited millions of years ago. The carbon in them was also absorbed by plants during photosynthesis, and represents locked up solar energy. But these deposits that took millions of years to accumulate are being dug up and released in a few decades, overwhelming the natural processes that recycle carbon.

The alternative sources of energy mentioned so far work – they are technically proven. There are also ideas for

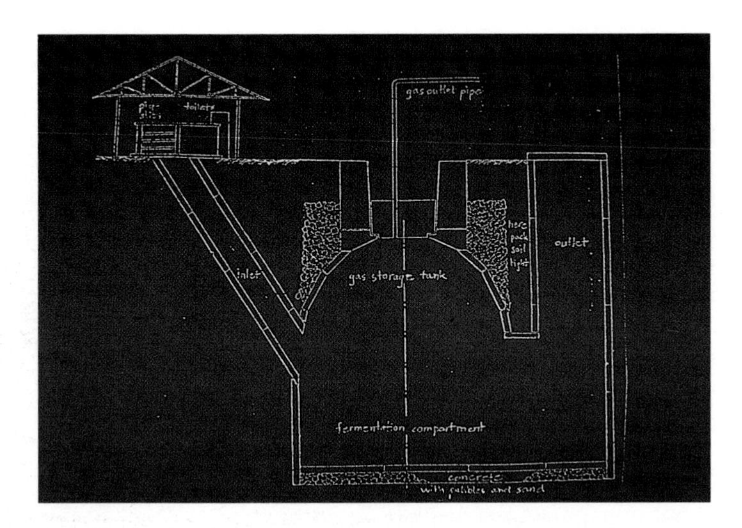

Biogas production in China

energy systems that have not yet been tested fully, but which have great potential. One possibility is to develop installations at sea that, in effect, draw heat from the contrast in temperature between the surface layers of the ocean and deeper layers. That is the kind of long-term scheme that might come into its own after the natural gas begins to run out. If other alternative energy sources are to make a real contribution in helping natural gas reduce our use of coal over the next 20 or 30 years, they must be developed on much larger scales, much more quickly than they have been in the past ten years. Indeed, development of these alternatives, wherever possible, is an immediate priority. But this will involve a shift in the emphasis of power generation from large, centralized plants to smaller, local systems – and also a shift in economic emphasis, perhaps with a tax on fossil fuels to make them more expensive and to encourage the use of alternatives. Judging by the reaction of the energy marketplace to oil price rises, it is estimated that a tax on fossil fuels equivalent to $10 on each ton of coal would reduce the rate at which carbon dioxide releases are growing by one percent per year. With energy demand today growing overall at less than two percent a year, this would be a very significant reduction.

There is one alternative to fossil fuel that we have not yet mentioned but is proven technologically, produces no carbon dioxide emissions, and fits in to the traditional infrastructure with large, centralized power plants. The questions surrounding this issue are so complex that it cannot be lumped in with a discussion of other non-fossil sources of energy, but warrants a full discussion of its own.

INTO THE FIRE?

Nuclear power works, and it does not produce emissions of greenhouse gases. But leaving aside debate about the desirability of and safety of nuclear power in its own right – for the record, we believe that the hazards are too great to justify more than a very limited use of nuclear energy – it is not a practicable proposition as a 'solution' to global warming.

Enthusiasts for nuclear power sometimes argue that the best way to reduce emissions of carbon dioxide is to replace

Nuclear reactor under construction in Mexico

existing fossil-fuel-fired power stations with nuclear plants, and to build only nuclear plants in the future. The political problems of such a policy are too obvious to need detailing. The practical difficulties of achieving such a goal have been spelled out most clearly by Bill Keepin and Gregory Kats, of the Rocky Mountain Institute, in the US. They deliberately used optimistic estimates of the cost of nuclear reactors and the speed with which they could be built, assuming that standardized reactors could be built at a cost of $1075 per kilowatt of installed generating power. British reactors have generally cost about twice that. They also assumed that the operating and maintenance costs would be half the current US figures. They then calculated how many nuclear power plants would have to be constructed between 1985 and 2025 to replace all existing coal-fired power stations and to cope with the expected increase in demand for coal-generated power in standard future energy scenarios (which assume at least a doubling in energy consumption over that period).

In a scenario in which no effort is made to achieve energy efficiency, it turns out that 'solving' the greenhouse problem by building nuclear power stations would require building 8000 large nuclear plants at a cost of more than $8 trillion at 1987 prices. And because oil and natural gas continue to be used in increasing quantities in this scenario, the carbon dioxide concentration in 2025 is still 65 percent higher than today: even for a more modest scenario in which energy demand reaches just over 21 terawatts by 2025, roughly twice the present-day value, in order to displace coal from the power generation scene we would have to bring 1000 megawatts of new nuclear power generating capacity on line every two and a half days between now and that year. 'Now' is the beginning of 1989; every month that passes without this programme beginning means a dozen extra power plants have to be added to the construction programme for future years. For comparison, in the peak years for the nuclear power industry, from 1970 to 1985, nuclear output worldwide increased at a rate of 1000 megawatts roughly every 24 days. In both scenarios, the massive commitment to nuclear power would reduce global warming due to carbon dioxide by only about 25 percent compared with what it would have been if none of these new nuclear plants had been built.

So, building nuclear power stations ten times faster than ever before would still only 'solve' a quarter of the carbon dioxide element of the greenhouse problem, in a business as usual scenario. And it would bring many other problems in its wake – not just the problems of safety, radioactivity and waste disposal, the danger of nuclear weapons proliferation, but those of the cost and upheaval of the construction programme itself, not to mention all the carbon dioxide pouring in to the air as a direct consequence of that construction activity.

The reason why even this massive effort to replace electricity generated by coal has such a small impact on the greenhouse effect is that electricity generation is itself only a small part of the greenhouse problem. Carbon dioxide from electricity generation is only part – a relatively small part – of total carbon dioxide emissions. And carbon dioxide itself is only half of the greenhouse problem. When Keepin and Kats say that a massive nuclear programme might cut the projected warming due to carbon dioxide by the year 2025 by 25 percent, that means holding the total warming due to all greenhouse emissions down by only about 12.5 percent compared with the level it would reach if uncontrolled.

Power stations fuelled by fossil energy account for only 15 percent of carbon dioxide emissions today. The rest comes from the destruction of the tropical forests, industry, transport, and fuel used in domestic households. Even in Britain, a coal and oil rich country, fossil fuel power stations make up only one-third of the nation's carbon dioxide emissions each year. Nuclear power is only useful for making electricity. It cannot fuel a truck, or an aircraft, or even a commercially viable ship (existing nuclear ships are run by governments, at a loss), nor is it much use for melting ore; and the prospect of a nuclear reactor in every home, rather than oil or gas fired central heating systems, is remote. In principle, it is possible to imagine a world run almost entirely on electricity, with most of that electricity produced by nuclear power. But even science fiction writers would not set such a scenario in the first half of the twenty-first century, and it has no relevance to the immediate practical problem of holding down emissions of greenhouse gases.

Something which sounds like science fiction and has recently been in the news is 'cold fusion' – the prospect of power generation from a previously unrecognized source. Even if this does prove more than a red herring, it will take many years to scale up 'test tube' experiments, producing a fraction of a watt of power, to power stations generating megawatts of energy. Whatever else it may be, cold fusion is not an immediate solution to the global warming. Indeed, it poses a threat. If large amounts of money and resources are devoted to attempts to develop this technique, fewer resources may be devoted to proven remedies such as increased energy efficiency.

The final nail in the coffin of nuclear power as an immediate solution to global warming also comes from the Rocky Mountain study. Even with all of their optimistic assumptions about the cost and efficiency of nuclear plants, it turns out that the cost of replacing each kilowatt of coal-fired electricity by a kilowatt of nuclear electricity is still seven times greater than the cost of saving a kilowatt of electricity by putting in to practice existing, proven energy efficiency programmes. Putting it another way, if you had a billion dollars to spend on tackling the greenhouse problem, you would do at least seven times more good by spending it on things like better insulation, more efficient vehicles, and less wasteful light bulbs than by building a nuclear power plant. What's more, you would achieve your objective much more quickly. As Keepin and Kats conclude: 'The fastest, least expensive, and above all, most effective response to greenhouse warming induced by fossil fuel combustion is to curtail the emission of carbon dioxide by improving the energy efficiency of the global economy.'

This is a fact which does not yet seem to have been taken on board by planners. For the next ten years at least, any time that a new power station, whether nuclear or not, is proposed in the industrialized world, the nation involved will be able to save money by not building that plant and by investing, instead, in energy efficiency. Of course, if you are a power supplier, who makes money by selling electricity, there is no incentive to do this. You want people to waste electricity, provided they are paying for it. It needs action at government level to achieve the savings, both in terms of money and carbon dioxide emissions, that are possible.

The fact that electricity generation is only part of the overall greenhouse problem is not all bad news. Since global warming is also caused by other activities we can tackle the problem in other ways, cutting back emissions of other greenhouse gases. In some cases, this is easier said than donc – but there is one obvious candidate for control, the CFCs.

CFCS AND OTHER GREENHOUSE GASES

Ten years ago, one of us wrote a book (*Future Worlds*, by John Gribbin, Plenum, 1981) which looked at the related problems of inequality and development, and how they might be overcome. In that book, the greenhouse effect was dismissed as a problem for the far future, on the basis that projections of the buildup of carbon dioxide in the air would 'only' bring a quadrupling of the natural concentration some time in the twenty-second century, and that this would lead to a warming of 'only' about 6°C. On that basis, the book urged the use of cheap, coal-fired power stations and industrial growth as the quickest aid to development – rather like the approach now being adopted in China. This was a mistake. It was made because, like almost every other study carried out in the late 1970s, the work on which *Future Worlds* was based did not consider the impact of greenhouse gases other than carbon dioxide. It is the buildup of these other gases in the atmosphere, on top of the rise in carbon dioxide levels, that makes the greenhouse problem so pressing. (Mind you, even a rise in temperatures of 6°C on a timescale of a couple of centuries now hardly seems any cause for complacency!) The corollary is that we can ease the pressure, making a dreadful prospect merely bad, by reducing the emissions of those other gases.

Having said that, there is very little prospect of halting the buildup of methane in the atmosphere, because methane production is so closely linked with agriculture. Such scope as there is of slowing the buildup should not, however, be ignored – especially since the appropriate action ties in, yet again, with policies that ought to be pursued for other reasons. Burning of tropical forests, for instance, releases methane, as well as carbon dioxide into the air. About 50 million tons of methane is produced this way each year, roughly one quarter as much as the combined release from paddy fields and cattle. Methane is also produced by the action of bacteria on household and other wastes at landfill sites. There, it builds up as a potentially lethal explosive hazard. Wastes might be treated in different ways, removing both the explosion threat and this particular contribution to the greenhouse effect. Existing landfill sites might be tapped, and

the gas being produced in them burnt. And, as we detail in the following chapter, we could just ensure that we produce less waste by recycling wherever possible.

Remember that each molecule of methane is 20 times more efficient at trapping heat than a single molecule of carbon dioxide. So, as we said earlier, burning methane, wherever and whenever it is getting in to the air, directly reduces the overall strength of the greenhouse effect even though carbon dioxide is produced. And the use of methane as a stove fuel would provide a reliable source of energy in regions now dependent on dwindling stocks of firewood.

Methane is also released from coal seams and oil wells when they are opened up for exploitation. Reducing the use of fossil fuel will also help to reduce methane emissions, and will reduce the production of nitrous oxide, another greenhouse gas that is formed when fuel burns. As we have seen, nitrous oxide is also produced from nitrogen fertilizers used in agriculture, and even if this use is reduced in some parts of the industrialized world, it will surely increase in the Third World over the next few decades. Nitrous oxide is also released from vehicle exhausts. The same vehicle exhausts, combined with other sources of pollution, cause acid rain and photochemical smog, and produce ozone, which is itself an undesirable greenhouse gas when it forms near the ground. Controls aimed at eliminating this kind of pollution will automatically slow the growth of the greenhouse effect, and that is another incentive, if one were needed, to develop more effective clean air technology.

But the greatest immediate easing of the greenhouse effect, giving us a breathing space of a full ten years before the anticipated doubling of the carbon dioxide equivalent of the air, would be a total and complete ban on the use of CFCs. Molecule for molecule, some CFCs are *10,000* times better at trapping heat than carbon dioxide. These gases are already the subject of concern because of the damage they do to the ozone layer, and there is an international agreement, the Montreal Protocol, to limit their emissions – although, as yet, this only calls for a cut of roughly 40 percent by the end of the twentieth century. In spite of its deficiencies, the Montreal Protocol is a landmark in the developing awareness of environmental issues, the first international, as opposed to bilateral, agreement to limit the damage we are doing to the atmosphere. It is important to make it work effectively, because it provides the archetype for other agreements which could soon be introduced to restrict the amount of carbon dioxide being emitted in to the air. The mechanism already exists for reviewing and revising the agreement, and it could easily be adapted to bring about a complete halt to the release of CFCs by signatory states by the year 1995. That would help the ozone layer, give us a little more time to cope with the growing greenhouse problem, and set the scene for controls on carbon dioxide itself. We will discuss the development of the Montreal Protocol further in Chapter 5.

Yet again, action required to get us out of the heat trap is action that ought to be taken anyway, and in this case it is being taken, albeit rather too slowly. But the place where all of these issues of humanity's relationship with the environment come together, all pointing to the same need for action, is in the forests. Not just the tropical forests, but forests wherever they occur.

SPEAKING FOR THE LAND

Tropical rainforests play a vital role as a major biospheric resource. Their preservation is a critical necessity for the continuation of life as we know it on planet Earth. Yet these forests are being destroyed at the rate of nearly one million acres a week. An area the size of Great Britain is destroyed every year. At this rate, there will be no tropical forest left by the middle of the next century; and as each tract of rainforest is destroyed, environmental stress and the impacts of greenhouse warming are intensified.

Extensive afforestation would take carbon dioxide out of the air and could preserve this resource. In a recent study, Gregg Marland of the Oak Ridge National Laboratory estimates that to offset the current injection of carbon due to fossil fuel combustion – about 5000 million metric tons of carbon a year – would necessitate the planting of seven million square kilometres of trees. Estimates vary. Marland's calculations were based on planting sycamore – inappropriate for the tropics where other species which take up carbon more effectively would be used. As only half of the carbon released into the atmosphere remains there, the rest being absorbed by the oceans and other sinks, and taking advantage of the higher efficiency of tropical species, the British ecologist, Norman Myers, considers that only three million square kilometres need be reforested.

But this is still a large area – about the same landmass as Zaire – leading some analysts to dismiss it as a pipedream. If we reduce this to the level of individual action, though, it means that each person on the planet would only be responsible for the planting of a small area, 25m by 25m square. This does not represent a large number of trees, and the goal comes within reach. Backed by governmental action, reforestation could play a major role in reducing the greenhouse effect.

But it would have to be reforestation of the right kind, the planting of suitable, indigenous trees in appropriate habitats. Much afforestation currently underway in the Third World involves the planting of eucalyptus, usually for commercial logging. Eucalyptus is chosen because it is relatively hardy and grows quickly. But it is extremely costly in terms of its water and nutrient requirements, rapidly draining the environment. And it is often a focal point of conflict between commercial gain and the needs of the local people. In one recent incident, villagers in Thailand cut through the stems of eucalyptus saplings, planted next to their village without consultation, below the ground; they left the young trees standing to sabotage the plantation and then laughed as the company men tried to figure out why the saplings weren't growing.

Some groups, keen to stress their environmental credentials, promote reforestation from an environmental angle though they are really concerned with commercial gain. A well-known multinational corporation, expanding into logging from its primary concern with fossil fuels, has tried to convince one Third World government that a eucalyptus plantation displacing 200,000 residents of a semi-forested area and providing employment for a mere 1000, will be 'good for the country'. Economically, perhaps, this may be true; but the 199,000 displaced people would probably disagree about the benefits they will receive. International agencies make the same mistake. The *Tropical Forests Action Plan*, promoted by the Food and Agricultural Organization in cooperation with the World Resources Institute, the World Bank and the UN Development Programme, is heavily based on commercial exploitation of reforested areas, blind to the interests of the local people. And the interests of the local inhabitants should be of paramount importance. They know what the land needs, and they know what *they* need.

All too many aid programmes simply make matters worse. The World Wide Fund for Nature (WWF) issued a detailed report on Japan's use of tropical timber in April 1989. Seventy percent of Japan's imports of tropical hard wood come from the Malaysian island of Borneo, 'threatening the continued existence of some of the most biologically valuable forest in the world.' WWF accuses the Japanese aid programme of subsidizing the destruction of the forests. The stated aim of aid projects may be to further development; in reality, concessionary loans and grants are opening up new areas to Japanese timber companies. 'The current destructive logging of tropical forests for Japanese markets should be rapidly phased out and replaced by sustainable practices,' warns WWF. Whether or not sustainable logging can be achieved on a commercial scale in the tropical forests is the subject of heated debate.

The destruction of the tropical rainforests is, sadly, an all too familiar story. But we can get a slightly different perspective on this destruction of the environment by looking, paradoxically, at a so-called developed nation and at how it treats a major part of its resources, and many of its own citizens, with the contempt that the colonial powers traditionally reserved for their colonies. That country is Canada, a nation still rich in forests that are being destroyed – largely to feed the hungry paper industry – in spite of strenuous efforts by the native people who live in those forests to preserve a way of life that is in harmony with the environment.

'There is no parliament for a wolf to sit in and make his point ... no parliament for a bear to sit. If the eagle could learn to speak English, he or she would say the same thing; and the caribou and the moose would say the same thing to the world. You've got to slow down industry, you've got to stop killing this land.'

Gary Potts of the Teme Anishnabai Indians

Canada is only marginally better than Central and South American nations at limiting the damage done to its forests. In the Third World, for every ten trees that are cut down, just one is planted. In Canada, for every ten trees that are felled, just three are planted. And Canada actually exports the machinery and skills used for forest clearance to the Third World. Members of the Indian nations that have occupied the Canadian forest land for at least 6000 years in a sustainable balance with nature see the impact of the modern despoliation of the homeland as 'like a pack of locusts, eating everything away.' These are the words of the chief of the Teme Anishnabai Indians, Gary Potts. He contrasts the attitude of modern society with that of his people, whose creed is that 'this land belongs to our people; some of us are living, some of us are dead, but most of us are not yet born; this land is not for sale.' And he eloquently describes how his people feel that it is their duty to speak for the land, because: 'There is no parliament for a wolf to sit in and make his point ... no parliament for a bear to sit. If the eagle could learn to speak English, he or she would say the same thing; and the caribou and the moose would say the same thing to the world. You've got to slow down industry, you've got to stop killing this land.'

Gary Potts

The greenhouse effect is just one consequence of this killing of the land. The Indian nations of Canada have already noticed the climate changing as a result. They have experienced the warming that most analysts only see in their computer databanks. Would the scientific community be more vociferous in its concern about global warming if it saw the effects firsthand rather than on the screen of a display unit? In northern Canada, there is less snow than there used to be, the Indians say, but day to day changes in the weather are more extreme. Instead of a cold winter with one thaw in spring, they now experience a thaw almost every month during the winter, with cold snaps in between. One day it might be slushy and thawing, the next temperatures may plummet to minus forty.

The Canadian example also shows, in a microcosm, how the crazy decisions that lead to such climatic upheaval and the killing of the land get made. 'When environments are destroyed, it's because people who don't have to live with the consequences of those decisions are actually making them, so they don't have an incentive to protect the environment,' says Pat Adams of Probe International. 'One solution is to return the decision-making power to the people who actually live in a community. We have to start at the grassroots level. We have to stop the wrong people from making decisions that affect other people's lives.' These decisions are made by people sitting in insulated office buildings far from the area affected. They do not feel the direct consequences of the decision to, say, open up a new region of forest to the machines that rip out trees wholesale and leave the land too damaged to recover. If the people who actually made such decisions were the ones who had to live with the consequences of their actions, then the decisions they made would be better for the local environment – and for the world as a whole. It is an obscene irony that the Canadian government, attempting to present an ecologically-aware face to the world, has taken an international lead in hosting conferences on ozone depletion and the greenhouse effect – while continuing to tear the heart out of its own forests.

'Agencies like the World Bank finance these mega projects, destroy rainforests, river valleys – the environments people depend on – and with impunity. They are unaccountable to us. They will not release documents to us. They will not allow public debate or public discussion of these projects. They are laws unto themselves.'
Pat Adams, Probe International

But *nobody* is really insulated from the greenhouse effect. Even decision-makers who sit in comfortable offices in capital cities are now being faced with the fact that the decisions they make *will* directly affect their own lifestyles, for the worse. It is a positive aspect of the greenhouse problem that it makes all these issues important to everyone. De-

cision-makers who felt no threat from the destruction of a forest ecology thousands of miles away can now see that their own homes are at risk from rising sea level, that their food supplies are threatened by drought. Perhaps they will start listening to the people at the grassroots who have been warning for years about the consequences of the environmental destruction they have seen around them. By looking at how political awareness of these problems has developed over recent years, we can see the way to make further progress, and how to take practical heed of the advice of those people who speak for the land.

But before we do, one final point. The steps outlined in this chapter represent a *first* step in dealing with global warming. For the most part, they represent technological shifts, adjustments in policy and changes in economic emphasis. In the longer term, we will have to get to grips with the root of the problem, taking hard decisions about the model of development that we need to achieve a sustainable future, about individual lifestyles and aspirations. Gary Potts, when he contrasts the brutal destruction of his homeland with the harmony of his Indian lifestyle, when he stresses the importance of ensuring his children, and their children, and their children's children inherit that homeland intact, is raising that issue. We should be listening, and listening hard. We tend to think of native peoples as backward, of ourselves as superior. We are wrong – there is much they can teach us about how to live our lives as promoters of life, rather than as agents of destruction.

'We are the leaders and representatives of many Penan communities in the Baram district of Sarawak. Altogether 105 of us were arrested between January 18–21 and 128 between November and January. The arrests relate to blockades of logging roads.

'For a long time we have suffered from the logging activities. Our forests resources are gone. Our food supply is reduced. Our river waters are polluted. Our rice farms and fruit trees are damaged. The wild animals have run away. Many complaints were sent by us to the district office, to the police, and other state authorities but they were not listened to. Some of us went to Kuala Lumpur to meet ministers who promised to help. But nothing happened. Our blockades which started in March 1987 were taken down by the authorities in October 1987. From October 1987 to August 1988 the situation worsened. The logging companies chopped the trees even faster. The state government created a new law making it an offence to blockade logging roads, to prevent us defending the forest.

'But many of our communities have suffered beyond endurance. We see our forests being logged away, our beautiful rivers made cloudy and poisonous by pollution. We suffer common problems of lack of food, health problems, some children have also died of hunger. The big soil erosion caused by logging silting our rivers – rivers that were once so deep, now made shallow by silt. As a result there was serious flooding and a lot of our food farms were damaged.

'So at the end of 1988 we were forced by our desperate situation to blockade again. But now the forest officials and police have come to our blockades and arrested us under the new forests law which forbids blockades.

'We ask that the authorities should not arrest any

more native people since the question of who has rights over customary land is not yet decided.

'We are the poorest people in the country. We are the victims of logging, of people from the big modern outside world who don't understand our system or our rights. We just want to live, like everybody else. We are used to the big forest and being free in a natural space. Being cramped in jail is very difficult for us. We love our land and forest very much, which our forefathers gave to us. We don't want to leave this land. Although a lot of forest where we live is already gone, yet we are willing to live here. Most of us have applied to the Government to grant us communal forest which is possible under state land laws. So much land is already given to the timber companies. We just want a little for ourselves and to know that it is protected from loggers and other people.

'We ask for help from people all over the world. We are people with a proud culture and a way of life that is based on our forest and land. Don't take our forest and culture and our dignity away. We thank everyone who thinks of us and helps us to solve our problems.'

Statement from six village chiefs and representatives and 14 Penan communities, April 1989

5

BACK TO GRASS ROOTS

Widespread public concern about the greenhouse effect dates from the late 1980s. But scientific interest in the phenomenon goes back more than 150 years. It was in 1827 that Count Fourier, the French mathematician who had been ennobled by Napoleon for his work in government, pointed out that the existence of a blanket of gases around the Earth keeps the surface of our planet warm. As far as we know, Fourier was also the first person to make the analogy with the action of a greenhouse – he referred to the way in which air inside a little wooden box with a glass lid gets hot when the box is left in sunlight. But in the 1820s nobody, not even Fourier, was worried that human activity might be strengthening the natural greenhouse effect.

Other nineteenth century sages dabbled with the science of the greenhouse effect. They included John Tyndall, in

Britain in the 1860s, and the American astronomer Samuel Pierpoint Langley. It was a Swede, Svante Arrhenius, who first calculated accurately the effect on global temperature of doubling the amount of carbon dioxide in the atmosphere. The calculation was published in 1896, and suggested that doubling the natural concentration of carbon dioxide in the atmosphere would increase global mean temperatures by about 5°C, very much in line with the latest estimates derived from sophisticated numerical modelling experiments. An American researcher, Thomas Chamberlain, also studied the phenomenon about the turn of the century, relating ice ages to carbon dioxide variations. Scientific interest in the subject then waxed and waned up until the 1930s, the first time that the greenhouse effect made a minor ripple in the world at large.

Many parts of the Northern Hemisphere warmed during the 1920s, reflecting the trend in temperature that affected much of the world at that time. Climatologists, reviewing the expanding amount of data becoming available through global coordination of the world weather network, spotted this trend and hypotheses concerning its possible cause multiplied. Several researchers drew attention to the possibility that the root cause might be the buildup of carbon dioxide in the air. The theory gained credibility when, in 1938, the British scientist Guy Stewart Callendar demonstrated that the amount of carbon dioxide in the atmosphere was, in fact, increasing. But there was no way to prove the link conclusively. For the first time, though, the notion of global warming attracted attention outside the scientific community.

Middle latitudes of the Northern Hemisphere cooled during subsequent decades and interest in the idea of global warming quickly faded. We now know that other parts of the world continued to warm but, during the 1940s and 1950s, climatologists were restricted in the geographical areas for which data were readily available. It is ironic that these were, by coincidence, the regions that were cooling. Atmospheric scientists concentrated on improving short-term weather forecasts – a subject which had more immediate benefits than the study of long-term climatic change.

RISING AWARENESS

Three developments triggered the current concern about global warming. First, the realization that humanity, through its disregard for the consequences of its actions, was creating a broad range of environmental problems, from local pollution through to the destruction of the ozone layer. The rise in ecological awareness reached its initial peak during the late 1960s, particularly in the US, resulting in widespread concern about the state of the planetary environment, population growth and resource availability. The mood of the times was reflected in the success of the

global bestseller *The Limits to Growth* and in the UN Conference on the Human Environment held in Stockholm in 1972. The UN's Environment Program (UNEP) was created as a result of this meeting. Widespread concern about environmental isses was now affecting the scientific community. In the case of the atmospheric sciences, it led to renewed interest in the subject of climatic change – and global warming.

All this fuelled the second development. Advances in understanding of the atmosphere, of the processes that shape climate and in computer technology resulted in rapid improvement in numerical models of the climate system. And the computer modellers were keen to experiment. Having modelled the present day, the climates of Mars and Venus and the ice ages, the greenhouse effect was an obvious next step. Regular measurements of the amount of carbon dioxide in the atmosphere were first made during the 1950s and, two decades on, the inexorable rise in carbon dioxide levels was clear. As study after study led to similar conclusions, a consensus gradually emerged within the scientific community that global warming did, indeed, pose a serious and credible threat.

A series of reviews of scientific understanding of the problem were undertaken during the late 1970s and 1980s, culminating in the most recent projects undertaken by the US Department of Energy and SCOPE, the Scientific Committee on Problems of the Environment. These two major reviews signalled scientist's concern about greenhouse warming to the world at large. In the foreword of the Department of Energy report, programme director Fred Koomanoff talked of the responsibility for the stewardship of the Earth that we all share and the importance of nurture 'rather than unrecognized neglect.'

The final, and perhaps most striking development, emerged in the mid-1970s. Soviet scientists, monitoring changes in the climate of their vast territory, reported that parts of Siberia were undergoing significant warming and cited the greenhouse effect as the probable cause. By the late 1970s, it had become clear that this warming was not restricted to Siberia and, during the mid-1980s, the first truly global estimate of surface temperature revealed the trend was worldwide. It became apparent that the cooling that had followed the 1940s and put a damper on northern interest in the greenhouse effect was merely a localized, short-term departure from the longer-term warming trend. As temperature records were broken during the early 1980s, the stage was set for the greenhouse effect to enter the arena of political debate with full force.

GREENHOUSE POLITICS

In June 1986, the US Environmental Protection Agency and UNEP organized a large scientific conference covering both ozone depletion and the greenhouse effect. The meeting crystallized the growing recognition that issues arising from pollution of the Earth's atmosphere are often closely inter-related: CFCs both destroy ozone and are greenhouse gases; conservation of fossil fuels would contribute to solving the problems of both acid rain and global warming. Three hundred politicians, decision-makers and scientists met for five days of collective discussion.

> 'Far-reaching impacts will be caused by global warming and sea level rise, which are becoming increasingly evident as a result of continuing growth of atmospheric concentrations of carbon dioxide and other greenhouse gases. Other major impacts are occurring from ozone layer depletion resulting in increased damage from ultra-violet radiation. The best predictions available indicate potentially severe economic and social dislocation for present and future generations, which will worsen international tensions and increase the risk of conflicts among and within nations. It is imperative to act now.'
>
> Conference Statement, *Our Changing Atmosphere: Implications for Global Security*, June 1988, Toronto, Canada

THE HEAT TRAP HAS BEEN SPRUNG. WE MUST ACT NOW TO LIMIT THE DAMAGE

Dear friend,

Suddenly the Greenhouse Effect is in the news. But the concept isn't new. In the last century scientists warned that the climate could be changed by increasing amounts of carbon dioxide in the atmosphere. Carbon dioxide and other greenhouse gases trap heat, so that it can't escape into space. The world's atmosphere is getting dangerously warmer. Governments are only now waking up to the scale of the potential disaster.

However, politicians rarely act until public pressure forces them. One of the most important jobs for Friends of the Earth is to mobilise the public to demand change.

It costs money to run our campaigns. We rely completely on donations from concerned people like you. We welcome any gift, no matter how small.

Please give whatever you can afford to help us protect our world.

Thank you for your help.

Jonathon Porritt

PS. If you can send £35 or more, we will send a full information pack plus a copy of **The Friends of the Earth Handbook.** This fascinating book covers many of the issues we are fighting on your behalf.

FRIENDS OF THE EARTH ARE FIGHTING ALL THESE ISSUES ON YOUR BEHALF...

RECYCLING
Our very first campaign was against throwaway bottles. There is still great scope for recycling glass, paper, cans and plastics.

AIR POLLUTION
Britain is Western Europe's largest emitter of sulphur dioxide. And we lag behind in dealing with poisonous exhausts from vehicles.

WATER POLLUTION
Much UK drinking water fails to meet EEC standards. The Irish Sea and the North Sea are filthy with chemicals and sewage.

SAVING THE RAIN FOREST
50 million acres of Brazil's forests gone in a year! Rainforest destruction could alter the world's climate.

CITIES FOR PEOPLE
Britain's transport policy has collapsed. The results of our transport research have been presented to a House of Commons Committee.

THE OZONE LAYER
CFC aerosols are banned in the USA, Canada, Norway, Sweden and Denmark. But our government didn't act, and it took a FoE campaign to prompt the phasing out of CFC aerosol use in the UK.

GREENHOUSE EFFECT
Many activities are affecting the atmosphere. In the next century we can expect significant sea level rises. Many fertile areas could become less productive.

ACID RAIN
Britain's trees are among the most damaged in Europe. Not enough is being done to clean sulphur emissions from our power stations.

SAFER ENERGY
After 30 years nuclear power produces only 4% of energy in the UK. But the dangers of radioactivity and nuclear waste will last for centuries.

...ALL THESE ISSUES AFFECT YOU.

Friends of the Earth, 26-28 Underwood Street, London N1 7JQ

PRINTED ON RECYCLED PAPER

In a keynote paper, Genady Golubev of UNEP warned that 'in the face of these global issues, the world's commitment to the environment is on trial. The issue, stripped to its essentials, is simple and unequivocal in its message, our legacy to the future is an environment less benign than that inherited from our forebears. The risks are sufficient to generate a collective concern that forebodes too much to wait out the quantifications of scientific research. Advocating patience is an invitation to be a spectator to our own destruction.'

The 1986 conference, drawing on the results of the SCOPE and Department of Energy reviews published about that time, represented a major landmark, bringing together scientists and politicians to discuss an issue which had previously been largely a matter of academic concern. It was followed, in 1987, by the report of the World Commission on Environment and Development, often referred to as the 'Brundtland Report' after its chair Gro Harlem Brundtland. The formation of the Brundtland Commission was described in Chapter 1. The Commission's report *Our Common Future* landed on the desks of governments worldwide. For many politicians, this was the first time that they had given serious consideration to the implications of environmental deterioration.

The Brundtland Report is an exhaustive summary of the myriad ways in which environmental degradation threatens the development of civilization; it also contains many suggested means of improving the situation. According to the Commission, sustainable development means 'promoting harmony among human beings and between humanity and nature' – words straight out of an ecologist manifesto! Observing that current national and international political institutions have failed to overcome the crises already confronting humanity, it called for:

- a political system that secures effective citizen participation in decision making;
- an economic system that is able to generate surpluses and technical knowledge on a self-reliant and sustained basis;
- a social system that provides for solutions for the tensions arising from disharmonious development;
- a production system that respects the obligation to preserve the ecological base for development;
- a technological system that can search continuously for new solutions;
- an international system that fosters sustainable patterns of trade and finance; and
- an administrative system that is flexible and has the capacity for self-correction.

A utopian dream? Perhaps. But the Brundtland Report clearly spells out the consequences should we choose to ignore these recommendations:

'In terms of absolute numbers there are more hungry people in the world than ever before, and their numbers are increasing. So are the numbers who cannot read or write, the numbers without safe water or safe and sound homes, and the numbers short of woodfuel with which to cook and warm themselves. The gap between rich and poor nations is widening – not shrinking – and there is little prospect, given present trends and institutional arrangements, that this process will be reversed.

'There are also environmental trends that threaten to radically alter the planet, that threaten the lives of many species upon it, including the human species. Each year another 6 million hectares of productive dryland turns into worthless desert ... More than 11 million hectares of forests are destroyed yearly, and this, over three decades, would equal an area about the size of India. Much of the forest is converted to low-grade farmland unable to support the farmers who settle it. In Europe, acid precipitation kills forests and lakes and damages the artistic and architectural heritage of nations; it may have acidified vast tracts of soil beyond reasonable hope of repair. The burning of fossil fuels puts into the atmosphere carbon dioxide, which is causing gradual global warming ... Other industrial gases threaten to deplete the

The loss of the world's rainforests . . . unless action is taken

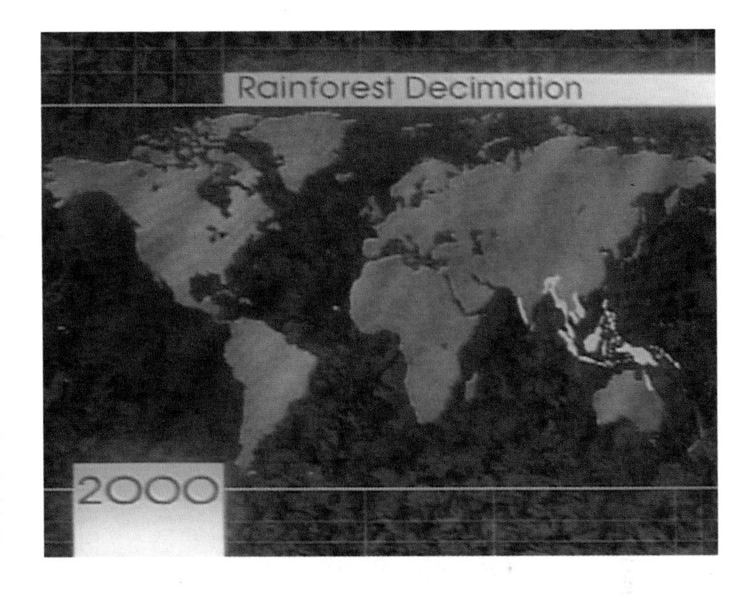

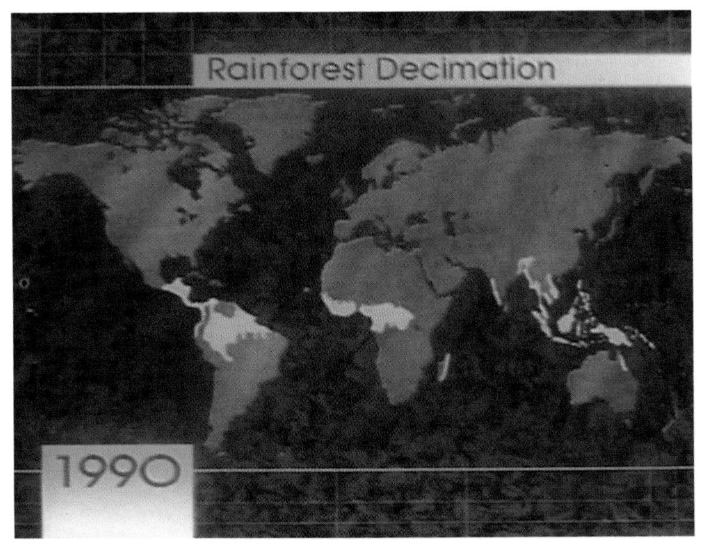

planet's protective ozone shield to such an extent that the number of human and animal cancers would rise sharply and the ocean's food chain will be disrupted.

'There has been a growing realization ... that it is impossible to separate economic development issues from environment issues; many forms of development erode the environmental resources upon which they must be based, and environmental degradation can undermine economic development. Poverty is a major cause and effect of global environmental problems. It is therefore futile to attempt to deal with environmental problems without a broader perspective that encompasses the factors underlying world poverty and international inequality.'

We have quoted at length from the Brundtland Report because we believe that this document contains the seeds of hope as far as the future is concerned. In linking environmental deterioration with the social and economic forces which determine development, *Our Common Future* provides a vision whose time has come.

'It is time that we realized that we all share a common future ... We need new concepts and new values to mobilize change ... We need to create more awareness and to mobilize people in all corners of the globe and in all walks of life. We need a sense of mission and to offer a common framework and a vision for a better future.'

Gro Harlem Brundtland

A CALL FOR ACTION

In retrospect, there could not have been a better time than June 1988 to arrange an international conference on global air pollution. Continuing drought in the heartland of the US had focused attention on the issue of global warming in the same way that the discovery of a hole in the Antarctic ozone layer had concentrated the minds of scientists, politicians and the world at large on the threat of ozone depletion. Drought in the Midwest is one of the consistent predictions emerging from computer model simulations of the greenhouse effect. In mid-June 1988, global warming hit the world's headlines when the US Congress investigated the problem. Climatologist Jim Hansen's statement at the hearings that 'it's time to stop waffling so much and say the evidence is pretty strong that the greenhouse effect is here' received global coverage.

The conference, 'The Changing Atmosphere: Implications for Global Security', was called by the Government of Canada and it was held in Toronto during the final days of June. Delegates included representatives of government, 15 international agencies, industry, and the environmental movement as well as scientists, economists and lawyers. Altogether, 330 people from 46 countries attended the meeting to discuss the implications of global warming and other global pollution problems, going beyond the scientific debate to draft

Global Warming Has Begun, Expert Tells Senate

Sharp Cut in Burning of Fossil Fuels Is Urged to Battle Shift in Climate

By PHILIP SHABECOFF
Special to The New York Times

WASHINGTON, June 23 — The earth has been warmer in the first five months of this year than on any comparable period since measurements began 120 years ago and the higher temperature can now be attributed to a long expected global warming trend linked to pollution, a space agency scientist reported today.

Until now, scientists have been cautious about attributing rising global temperatures of recent years to the predicted global warming caused by pollutants in the atmosphere, known as the "greenhouse effect." But today Dr. James E. Hansen of the National Aeronautics and Space Administration told a Congressional committee that it was 99 percent certain that the warming trend was not a natural variation but was caused by a buildup of carbon dioxide and other artificial gases in the atmosphere.

Global Warming: Greenhouse Effect?

Average globe temperatures through the first five months of 1988. As a baseline, scientists use the global average from 1950 to 1980.

Source: James E. Hansen and Sergei Lebedeff

The New York Times/June 24, 1988

Source: The New York Times/*June 24, 1988*

policy recommendations. The participants concluded the problem was so pressing that 'governments, the United Nations, and its specialized agencies, industry, educational institutions, non-governmental organizations and individuals should take specific actions to reduce the impending crisis caused by pollution of the atmosphere.' Moreover, 'no country can tackle this problem in isolation. International cooperation in the management and monitoring of, and research on, this shared resource – the atmosphere – is essential.'

The conference participants made several specific recommendations, listed at the end of this chapter. Among the most important, and pressing, they urged governments to strengthen the 1987 Montreal Protocol on control of CFC emissions in order to eliminate releases by the year 2000 and to make a commitment to reduce carbon dioxide emissions by one-fifth, starting from 1988 levels, by the year 2005. The carbon dioxide targets are not too ambitious and could easily be met in the industrialized world through improved energy efficiency and switching to non-fossil fuels. The conference also called for reductions in releases of the other greenhouse gases, and went on to outline a programme for further action.

This programme begins with the development of a comprehensive global convention for protocols on the protection of the atmosphere – making the greenhouse effect and related problems the subject of genuine international law, not merely relying on the goodwill of individual governments. Further research will be necessary to reduce uncertainties in the forecasts and to support the identification of appropriate policies. Most research on global warming has been undertaken from an academic perspective rather than within a tight political framework. The conference strongly recommended that the industrialized nations should actively seek ways to assist Third World nations to play their part in limiting the problem. The immediate priority should be to protect the tropical forests and promote reforestation. It was suggested that a trust fund should be set up by the rich nations to help the poorer countries of the Third World in protecting their environment, controlling pollution and preserving their dwindling resource base. The importance of involving Third World nations in monitoring and research programmes was stressed. Last but not least, the conference suggested that increased funding to non-governmental and educational establishments would be an effective means of raising awareness of the problem.

> 'Changes must be made if disastrous mistakes are to be avoided, but we also believe that it is possible to make these changes. Human resources, knowledge and capabilities have never been greater. We have the power to create a future which is more prosperous, more just and more secure for all.'
>
> Gro Harlem Brundtland

The Toronto Statement, as it has become known, is not the definitive blueprint for action to curb global warming. It has little to say about the social and economic factors which lead to environmental degradation in the Third World – it tends towards a legislative response to the problem rather than in the direction of social change. Nevertheless, it does represent a very sensible starting point. The most important feature of the conference, though, was that it marked a shift in emphasis in the approach to the issue. In the early 1980s, the *scientific* uncertainties in the projections and scenarios were always to the fore. The attitude of the politicians was, more or less, that as long as the scientists couldn't agree, there was no point in taking action. As the scientific uncertainties have been gradually resolved and the scientific community has reached a broad consensus, remaining doubts concern humanity's ability to respond effectively to the threat. *Political* uncertainty has replaced scientific uncertainty at the forefront of the debate. Can we reach international agreement to limit the impact of global warming? Or will we have to learn to live with the consequences of our complacency? In fact, there are encouraging signs that attitudes are changing, that we may be able to slow down the change in climate.

'What's the use of having developed a science well enough to make predictions, if in the end all we're willing to do is stand around and wait for them to come true?'

Sherwood Rowland, University of California at Irvine

THE GLOBAL GREENHOUSE NETWORK

As politicians, industrialists and the proverbial person in the street have adopted 'greener' attitudes over the past few years, environmentalist organizations such as Friends of the Earth and Greenpeace can be forgiven for allowing themselves a few moments of self-congratulation. It may have been a long, hard struggle, but there is no doubt that their incessant lobbying and campaigning is at long last paying off. On the other hand, there is still a long way to go. And the sudden surge of interest in global warming took many groups by surprise. They were left wondering just how to deal with the greenhouse effect as a campaigning issue – an issue that was truly global and warranted a global campaign, an issue that covered more fronts, environmental, social and political, than any other. It wasn't even clear what policy goals should be set. Was demanding a 20 percent reduction in carbon dioxide emissions by the year 2005 too modest? Was it over-ambitious?

'The environment is the business of everybody, development is the business of everybody, life and living is the business of everybody. I think the solution will be found in encouraging mass environmental literacy so there can be democratic and literate decisions, because if decisions are taken by a few without the incorporation of the opinion of the masses, the non-governmental organizations especially included, the likelihood is that the situations will not succeed. They will be imposed from above, the people will not respond positively to them, and the project is lost before it is launched.'

Joseph Ouma, Moi University

By late 1988, the major environmental organizations all had global warming projects underway and, in the autumn of that year, the first global campaign was launched. The *Global Greenhouse Network* was formed in October 1988 at a three-day meeting in Washington DC when key activists from 35 nations announced plans for a massive global grassroots mobilization on the greenhouse issue. The conference was sponsored by the Washington-based Foundation on Economic Trends, whose director, Jeremy Rifkin, has been in the vanguard of many campaigns.

'The greenhouse effect is an opportunity. It's an opportunity for us to make a world that's more efficient and less polluted. And, at the same time, to put at the very top priority the single most important issue – environmentally sustainable development. If it takes the threat of unprecedented climate change to do that, let's take that threat and solve two problems at once, environment and development. I think we can do it. And that's the good side of the greenhouse effect. The fear of warming may well have a very nice silver lining.'

Steve Schneider, National Center for Atmospheric Research

The meeting brought together 90 delegates from around the world representing a broad range of concerns. The participants represented environmental organizations, agriculture and farm groups, the scientific community, religious groups, food and consumer organizations, animal welfare groups, health advocates, wilderness preservation groups, and the entertainment community. In a statement prepared during the conference, the participants called for a range of policies to be adopted to combat the greenhouse effect and committed themselves to a decade of grassroots action. The Network is currently active in a number of areas. It is campaigning for more stringent controls on CFC production and use; developing a consumer guide concerning individual responses to the greenhouse problem; coordinating a 'Mayor's Campaign', through which cities and local communities link up and commit themselves to policies directed towards limiting global warming such as tree-planting and reducing urban pollution; producing a briefing document on global warming and the Third World; liaising with the entertainment community (film, art and music) to utilize their talents and celebrity status; and planning a global 'Reforest the Earth' project to promote individual action to combat the greenhouse effect.

The Network includes representatives from both North and South, East and West, providing a more amicable forum for the discussion of sensitive issues than does the political arena. And, in fact, there was vigorous and constructive debate during the Washington meeting between the delegates from the industrialized world and from the South. As the North has produced the bulk of the pollution already in the atmosphere, should it not compensate the Third World for causing global warming? Would it be reasonable

Who is to Blame?

We all are, one way or another. In industrial countries, consumer demand for rainforest resources and inappropriate aid programmes compound the problems caused by entrenched poverty, unequal land distribution and rising populations in the Tropics.

Every year 200,000 square kilometres of tropical forests are cleared or seriously degraded. Or over 1 million acres every week. Or 100 acres every minute.

Red Eye Tree Frog Ecology Pictures, M&T Fogden

The greatest threats are the spread of non-sustainable farming and logging. Millions of poor people in tropical countries clear the forests to grow food to eat and sell. In Latin America, cattle ranchers burn the forest to raise cheap beef for export. In West Africa and South-East Asia, commercial loggers cut down the world's best hardwoods such as teak and mahogany – and they rarely replant. The governments of developing countries construct huge projects like highways, hydroelectric dams and mines through formerly untouched forests, usually with massive foreign aid from industrial countries like Britain, the USA and Japan.

Logging Operations WWF

By 1980, some 40% of all tropical forests had gone.

In Africa, almost 6 million acres of tropical dry forests were destroyed each year between 1980 and 1985; in Asia, more than 4 million acres of rainforest were destroyed annually between 1976 and 1980.

In the last 30 years, Central America has lost almost two-thirds of its rainforests to cattle ranching.

Commercial logging devastates 12·5 million acres of tropical forests annually.

RAINFOREST

Rainforest and Reef Australian Conservation Foundation

Campaign to Save Tropical Rainforests

FRIENDS OF THE EARTH

26-28 Underwood Street, London N1 7JQ. 01-490 1555

for Third World nations to insist on debt write-off before they consider the adoption of expensive pollution control technology? A number of issues raised at the meeting have since hit the headlines. For example, Brazil has recently suggested that it might sue the industrialized world as it regards the high incidence of skin cancer amongst one section of its population to be the result of ozone depletion! The industrialized nations are, of course, responsible for nearly all releases of ozone-depleting chemicals.

CAN POLITICIANS CHANGE THEIR SPOTS?

As environmental concern spread amongst the general population, it was hardly surprising that politicians should sense the changing climate of opinion and stake their claim to green credibility. Some were genuine, others can be viewed with scepticism. But whatever the depth of this new concern, there is no doubt that the environment, with global warming to the fore, is now firmly on the political agenda.

Gorbachev addressing the United Nations

Mikhail Gorbachev, in a plenary address to the Peace Forum in Moscow in February 1987, called for a rapid end to the arms race so that nations could cooperate freely in dealing with the real threat to global security – the threat to the planetary environment. Environmental issues played a crucial role in the 1988 presidential elections in the US with all leading candidates stressing their environmental credentials, and the poverty of their opponent's claims. And, in a speech to the Royal Society in September 1988 which reverberated around the world, UK Prime Minister Margaret Thatcher executed an astonishing 'U-turn', stressing her belief that the environment should be nurtured and protected. 'It is possible,' she said, 'that with all these enormous changes – population, agricultural, use of fossil fuels – concentrated in such a short period of time, we have unwittingly begun a massive experiment with the system of the planet itself.' The speech represented a major shift in position; the UK is considered the 'dirty man of Europe' because of its lamentable record on the protection of the environment. Whether the result of genuine concern, vote-chasing or a desire not to be left out of the latest international debate, the fact that the environment has attracted the attention of the world's leading politicians is a very positive development. But will words be translated into action?

WHAT IS BEING DONE?

The international response to global warming will be marshalled by the United Nations Environment Program (UNEP), in its role as caretaker of the global environment. UNEP was formed in 1972 as a result of the Stockholm Conference. Based in Nairobi, Kenya, UNEP's role is threefold: to monitor trends in the environment, to stimulate relevant research, and to assist with the development of appropriate policy responses. The way in which it carries out its responsibilities can be seen clearly in its sterling work on ozone depletion.

UNEP became involved in the ozone depletion issue not long after its formation and, during the late 1970s, sponsored a series of studies, reviews and conferences on the threat posed by rising atmospheric concentrations of the ozone-depleters, the chlorofluorocarbons. Convinced that the scientific case was strong enough to warrant immediate action to curb releases of these chemicals, UNEP called on governments to reduce production and emissions as early as April 1980. Few responded at that time, but UNEP persisted and moved on to the next stage of their mandate – from science to the formulation of policy. The organization established a working group composed of both legal and scientific experts to draft a convention on which international law could be based and which would guarantee the safety of the ozone layer.

The eventual outcome was the Vienna Convention, signed by 20 nations in March 1985. It dealt in generalities and represented little more than an agreement in principle; but it set the stage for later developments. The signatories recognized that they had an obligation to control any activities that might harm the ozone layer and committed themselves to more research. Events then began to move swiftly. Two months later, Joe Farman and colleagues from the British Antarctic Survey published their evidence of a pronounced 'hole' in the ozone layer over Antarctica, catching the imagination and translating academic into public concern. At the same time, lines were being drawn on the political arena. Some nations, such as the US, Canada and Switzerland were pressing for an immediate ban on all uses of CFCs in spray-cans. Finland, Norway and Sweden wanted a more general ban. The UK and France felt that control was premature and that more research was needed – 'more interested in short-term profits than in the protection of the environment for future generations' was the verdict of Richard Benedick, US Deputy Assistant Secretary for the Environment. In fact, further research was revealing that the problem was even more serious than originally envisaged and that the need for action was becoming ever more urgent.

As the signatories of the Vienna Convention sat down to hammer out the details of the control measures which would be attached to it the debate became more and more acrimonious. The European Community wanted to present a united front in the negotiations but Britain refused to toe the line. So separate discussions had to take place within Europe before Europe could argue its position in the wider debate. The major CFC producers and users in the Third World – India, China and South Korea – refused to take any part in the discussions. Nevertheless, on 16th September 1987, after compromise and trade-off, the Montreal Protocol was signed by the representatives of 27 nations. It represented a major achievement for UNEP – the first international agreement to curb global air pollution.

What can we learn from this process about any future attempt to limit global warming through international action? Well, the most important point is that reaching international agreement can be a lengthy affair. The Montreal Protocol won't result in actual reductions in emissions for some years yet – we may see results *ten to fifteen years on* from the moment back in 1980 when UNEP decided that the scientific evidence was sufficient to warrant governmental action. Emissions of carbon dioxide and other greenhouse gases are related to far more fundamental aspects of life than are releases of CFCs; they are of far greater economic significance. It can be anticipated that the negotiations on limiting global warming will be far more acrimonious than those that accompanied the development of the Montreal Protocol, and that the timescale for reaching agreement could well be longer.

More can be learnt from the Montreal Protocol. The fact that it exists represents a major achievement. Nevertheless, it is totally inadequate. Compromises made during the negotiations resulted in targets well below those recommended by the technical experts and, shortly after the Protocol was signed, further evidence from the Antarctic demonstrated that even those scientific recommendations were not tight enough. The Protocol contains arrangements for regular review in the light of scientific developments and it now seems likely that it will be strengthened over the course of the next couple of years – but only after another round of political in-fighting and bickering has run its course.

The political manoeuvring has already begun. The British government, keen to demonstrate its new-found care for the environment, called an international meeting in London in March 1989 to discuss the revision of the Montreal Protocol. Attended by dignitaries from 124 nations, and accompanied by a flurry of diplomatic activity, the meeting had more in common with discussions on arms control and economic policy than it had with previous meetings on environmental affairs. And one aspect of the London debate presages what may prove the major stumbling block when it comes to reaching agreement on the greenhouse effect.

FAIR PLAY FOR THE THIRD WORLD

The Montreal Protocol attempted to favour the Third World in order to allow them to develop further their use of CFCs while the industrialized world cuts back. It did this by allowing low-CFC consumers to continue expanding production and use for some time after controls were to be applied by high-CFC consuming nations – a delay of a few years. Again, in recognizing that it is unjust to ask the Third World to forego the benefits of cheap refrigerants and so on which the North has already taken advantage of, the Montreal Protocol sets an important precedent. But the measures do not go far enough.

The dispensations given to low-CFC consumers have not been sufficient to convince nations such as China and India that they should sign the agreement. As China, for example, has stated that it intends to expand CFC production many times over coming years, its participation in the Protocol is considered a crucial factor. But you can see why the Third World is not convinced by the concessions available at present. If they follow the Protocol, they will be able to make use of CFCs for a few more years but ultimately they will have to switch to alternatives. They will be left with CFC technology which is obsolete. The North will have developed alternatives by that time and the Third World will be forced to pay high prices for the new production technology and for the chemicals themselves. They will be left at a pronounced disadvantage.

There is a related problem. Much was made at the London meeting of the *future* contribution of the Third World to global pollution. And there is resentment in the Third World that it is being held responsible for its future actions when, at present, it contributes relatively little to the problem. At this time, China currently produces 22,000 tonnes of CFCs each year and the US 300,000 tonnes. With a population of 1.1 billion people, almost a quarter of the world's population, China is only responsible for two percent of the world's releases of CFCs and related gases. That contribution will increase and it is important that Third World nations cooperate in controlling emissions. But there is a justifiable feeling that the industrialized world should concentrate on setting its own house in order rather than focusing attention on the *potential* contribution of the Third World. It is almost as though the North is trying to divert attention away from the sins of its past.

Is there an acceptable alternative to the provisions of the Montreal Protocol? The most sensible course of action seems to be to set the same control targets for both North and South but, in recognition of the hardship this will cause the poorer nations and the responsibility of the industrialized world in creating the problem in the first place, to establish a formal system of aid and assistance so that the Third World can meet stringent goals without hampering its own prospects for development. This aid and assistance could be in the form of finance, debt write-off, transfer of technology or fair trade agreements. But it must be formally written into the revised Protocol as a necessary precondition if Third World nations are to subscribe. In fact, China suggested at the London meeting that a trust fund of this nature should be established. But the proposal met with a mixed response. The US representatives greeted it favourably but Margaret Thatcher refused to accept that this form of assistance was necessary.

These issues will have to be dealt with in the context of negotiations over global warming. But the stakes will be higher. For example, it is in the Third World that the consumption of cheap fossil fuels is likely to grow at the fastest rate. Cheap energy is crucial if development is to occur. It is extremely unlikely that

Coal-fired power station in China

the Third World will jeopardize the process of development without the guarantee of aid and assistance from those nations they regard as primarily responsible for the problem. Brazil, reacting defensively to international criticism of the rate of destruction of the Amazonian forest, has made its position clear. Put rather bluntly, their argument goes like this. Why should we not make use of our resources in the way we see fit? After all, you, the industrialized world, have already destroyed your forests in the course of your own development. Why should we not follow the same path?

It seems unlikely that the Third World will cooperate fully unless the North makes concessions, recognizing that, to date, it is the main cause of global warming. Writing off debt would be a promising start; not only as a gesture of good faith but also to relieve the economic pressure on Southern nations so that they can afford to preserve forests and control pollution. Debt-for-nature schemes, whereby debt is 'sold' back to the borrower at a low price in exchange for a commitment to protect the environment, represent one means of coupling debt relief with ecological protection. Conservation International and the World Wide Fund for Nature (WWF) have been at the forefront of these initiatives. If implemented with due regard for social justice, and not as a means of commercial gain, they are an important step in the right direction. But more will be needed. 'If we want [the developing] economies to choose paths that value our shared heritage, we're going to have to change our trade policies as well as our aid policies,' says Irving Mintzer. The World Commission on Environment and Development argues strongly that fairer terms of trade will be essential if the rate of environmental destruction in the Third World is to be reduced. Amongst other things, this means fair prices for Third World commodities and an end to the preferential subsidies and trade barriers that favour Northern producers. It would represent a radical change – a change for the better – in the world's economic system.

Carajas Iron Ore Project, Amazonia

THE INTERGOVERNMENTAL PANEL

Whatever the difficulties ahead, the process of reaching international agreement to limit emissions of the greenhouse gases is underway. UNEP, in collaboration with the World Meteorological Organization, has formed the Intergovernmental Panel on Climate Change (IPCC). The purpose of this body is to review the latest scientific evidence concerning the physical science of the problem and its potential impacts. But its brief goes further. It will also consider appropriate policy responses on the part of the international community. The IPCC represents an extremely important development – the point where, for UNEP, global warming becomes a matter of international political concern. 'What is extremely interesting,' comments Mostafa Tolba, head of UNEP, 'is that for the first time

Mostafa Tolba

governments are now asking us to move on the issue of climate change – usually it's us pushing governments to get things done.'

The work of the IPCC is being handled by three groups. The UK is leading a working group on the basic physical science. The USSR is coordinating the impacts working group. The policy group is led by the US. The Panel began work in early 1989 and it will report to the World Climate Conference which is being held in 1990. It will be a hectic 18-month period for the working group participants. UNEP hopes that the conclusions of the IPCC will lay the basis for a future protocol on releases of the greenhouse gases, perhaps the Law of the Air advocated by the Toronto Conference.

But there will still be a long way to go. For example, international law is often disregarded. Nicaragua is still waiting for compensation from the US because of its illegal support of the counter-revolutionary 'contra' army funded and organized by the US government. This compensation was ordered by the International Court of Justice in the Hague in June 1986 but the American government simply refused to accept its jurisdiction. There is no international body empowered to enforce any future 'Law of the Air'. Aware of this lack, a group of nations, led by Norway and France, met in early 1989 to discuss how it might be remedied. Perhaps we need an Environmental Security Council to ensure the ecological integrity of the planet – a parallel to the Security Council of the UN, though without the undue power granted to the superpowers in that forum.

Moreover, the IPCC has a strong technical focus. This is understandable as the threat of global warming has emerged from the hard science of climate modelling. But as we have seen throughout this book, the greenhouse effect is not solely a scientific issue. It is also a matter of social justice, of economics, of politics. Its causes are deeply rooted in the process of development. Dealing with global warming will not simply be a case of the application of technology, of legislation to reduce pollution, of minor policy adjustments here and there. Dealing with the problem must involve deep questioning of the attitudes that have brought us to the current threshold, a thorough evaluation of models of development, of aspirations and of lifestyles. And this is beyond the remit of the IPCC.

Quite how this aspect of our response to global warming is to develop is not clear. *Our Common Future* provides an excellent starting point and the WCED has set in motion a number of initiatives to pursue its recommendations. In Norway, for example, the government has invested in numerous projects concerning sustainable development which are being undertaken by schools, universities, campaigning groups, development organizations, and so on. Each government department has been asked to provide a comprehensive assessment of the implications of the WCED report for its policies. But even *Our Common Future* stops short of questioning the dominant model of development. Assuming growth *can* be achieved without cost to the environment, it presents

no radical alternatives but concentrates on more effective means of achieving current goals.

Vandana Shiva of the Research Foundation for Science and Ecology in Dehra Dun, India, accepts that *Our Common Future* has much to offer but feels that 'at the moment we have a common future only in extinction. In the immediate five to ten years people in agribusiness do not have a common future with the peasant in Asia or Africa who is dying. A peasant in India about to be displaced by a dam funded by the World Bank does not have a common future with the fellow in Washington with their thousands of dollars salaries. But the theme of a common future has been used to allow the dominant powers and the industrial interests they serve to say: "Listen, there's a common global crisis – we've got the funds, we've got the technology, we've got the answers." It is a way to appropriate the terms of ecological resistance and so to regain control.'

Many would say that changes in attitudes, in aspirations, in lifestyles are a matter of individual concern, not the subject for grand international committees. And individual involvement is the topic that we shall end on. It will take many years for governments to reach agreement, for control measures to be adopted and implemented, but action at the level of the individual – you, the reader – will show immediate results. It will also create a climate of opinion which will galvanize governments, which will encourage industrialists to bring new, environmentally-sound products onto the market, and it may result in the fundamental change in attitude which is necessary if we are to deal effectively with global warming.

'It is entirely within the power of humanity to close the gap between rich and poor and to reduce the human population size to a level at which all people could lead a decent life without degrading the ecosphere. A transition to living primarily on income can be made; agricultural systems can be designed that would be highly productive and would also help support the natural ecosystems in which they are embedded and on which they depend. Societies can turn their backs on racism, sexism, gross economic inequality and, above all, war as a mode of problem-solving.... There are, in short, no insuperable barriers to creating a peaceful Earth in which Homo Sapiens leads a rich existence without overstressing the natural systems that support human life.'

Anne and Paul Ehrlich, *Earth*

FROM INDIA TO THE AMAZON

It is appropriate that we begin this discussion of the power of individual action with the tribal people who are still at one with their environment. 'This land is our cradle,' says Gary Potts of the Teme Anishnabai Indians. 'It's hard to separate it. It's a part of our whole. The land is our resting place. I just lose words . . . there are no words to express what it is to live within the womb of our mother.' Viewing the environmental degradation around them, it is hardly surprising that these people have taken action.

Reni is a village in the Chamoli district of northern India. In 1974, commercial loggers moved into the area, threatening not just the environment but the very way of life of the villagers. The women of the village took action, wrapping their arms around the trees to protect them from the chainsaws. Their defence was successful, logging was banned in the area, and the Chipko movement – 'the movement to hug' – was born. Tree-hugging has been adopted as a tactic in many parts of the Third World and the Chipko movement now runs reforestation projects throughout northern India.

Nicholas Hildyard of the *Ecologist* describes the scene as Indians and environmentalists from all over Brazil met at Altimira to protest at plans to build at massive dam – since abandoned – in the Amazonian rainforest:

'The Indians sat motionless on the concrete floor, as they had done for almost two hours, listening patiently as José Antonio Muniz Lopes, the chief engineer of Brazil's electricity conglomerate Eletronorte, smoothly outlined his case for building a massive hydroelectric scheme on the Xingu river. "The dams will be in your interest. Nor is it certain that they will even go ahead: they are still very much in the planning stage. We still have many studies to complete and only when they are finished will we make the decision whether or not to proceed with the project. But if they do, they will bring progress."

'The 600 Indians rose as one, raising their arrows and clubs in protest and chanting their disapproval. A woman, streaked in warpaint, strode to the dias, brandishing a machete and cutting the air to emphasize her points. Just inches away from Muniz, she brought down the machete in a graceful but swift arc, stopping the blade a hair's breadth from his shoulder blade. Muniz sat impassively as she ritually pressed the flat of the blade against his cheeks.

'"You are a liar. We don't need electricity. Electricity won't give us food. We need the rivers to flow freely; our future

Kaiapo Indian from Amazonia

depends on it. We need our forests to hunt and gather in. We don't want your dam. Everything you tell us is a lie."

'The tension was palpable. Around the edge of the hall, the small detachment of military police, who had been called in after shots were fired near where the Indians were encamped, fingered their holsters. A few Indian warriors, wary of possible violence, deftly placed their arrows in their bows. For a moment it looked as if the worst was about to happen.

'Muniz continued. "The dam project will only flood a small amount of Indian land. At most, 216 people will have to be resettled. It is a small price to pay for the economic benefits which will come to the region."

'But the Indians would have none of it. "You say you're conducting exhaustive studies, but you haven't asked us what we think. In ten years time, you might come and ask us our opinion, but in ten years time, there will be no Indians left. All you think about is economics. You don't think about us." "We are not here for show. We are not just dressed up for the cameras. We don't want this dam and we won't have this dam. You say you'll conduct more studies, but you don't need any more studies. The dam will be a disaster; you only have to look at the others you have built." "The government is out to integrate us. But how will that help us? Look at this town: it's a miserable place. Conditions are terrible, and the people live a miserable life. Is this what you are offering us? Is this progress? Why don't you spend your money to improve conditions here? When will you finally learn that your dams don't help anyone but the rich? Why won't you listen to us? We have been here for thousands of years. We will teach you how to live properly. Don't talk to us about relieving our "poverty". We are not poor. We are the richest people in Brazil. We are not wretched. We are Indians."'

Words such as these reflect a dignity and a sense of value that we, in the industrialized world, have all but lost.

THE PLANET IS IN OUR HANDS

Faced with the prospect of global warming, cynical about the short-sighted attitudes of politicians, dwarfed by the power of the large multinational corporations, it is understandable that many people feel little but despair. Global warming is just one more issue that they would rather not think about as there seems little that they can achieve. What can we do to save the planet in the face of what appear to be insurmountable problems? How can my actions prevent a rainforest being destroyed in Brazil, stop the construction of coal-fired power stations, ease the burden of Third World debt? In fact, there is a *great* deal that we can all do. The reason is simple. Politicians want to get elected. Manufacturing companies want to sell their goods. And the power of the individual to influence the actions of both government and industry is nowhere better seen than in the recent history of the threat to the ozone layer posed by CFCs.

What has become known as the 'first ozone war' was largely fought in the US in the 1970s, and chiefly concerned the CFCs used as propellants in aerosol spray cans. When scientists first identified the possibility that chlorine from CFCs might reach the upper atmosphere and destroy ozone, allowing cancer-causing ultraviolet radiation to reach the ground in increasing quantities, the reaction of industry was that this was 'just a theory' and that no action need be taken to alter their products until years of observations had tested the idea. The scientists responded that by the time conclusive evidence was available, CFC levels in the atmosphere would be so high that the damage would have been done. Similar arguments have been voiced concerning global warming. While the debate raged during the mid-1970s, the public voted with their money, at least in the US. Sales of all aerosol sprays – whether they contained CFCs or not – collapsed, and the manufacturers of products that did not contain CFCs began to publicize the fact, picking up a growing share of the market.

Although industry complained, an important principle was, if not established, at least implicit in this development. Reversing what we think of as the law of natural justice, products released into the atmosphere came to be regarded as 'guilty until proven innocent', and not the other way round. Following this principle, manufacturers should be constrained to prove conclusively that their products are environmentally safe *before* they can be used – and not be allowed to leave the burden of proof of guilt to the slow progress of scientific understanding and experimentation on the real world. With consumers shunning sprays containing CFCs, and voters, those same consumers, vociferously calling for political action, the US government introduced legislation severely restricting the use of CFCs in aerosol cans in the late 1970s.

For a time, this made a major contribution to the global control of CFC emissions, because the US was then the world's biggest user and spray cans were the main application. But, by the 1980s, CFCs were being increasingly used in other applications and other countries continued to use them in aerosol cans. In fact, many of the same manufacturers who had switched to safer alternatives in the US sold the same products propelled by CFCs abroad. This hypocrisy, leading to double and triple standards depending on local laws and consumer attitudes, is something that the multinational corporations should not be allowed to get away with.

The 'second ozone war' was triggered by the discovery of the hole in the ozone layer over Antarctica – a slice out of the ozone shield as deep as Mount Everest is tall, covering an area the size of the contiguous United States. This led to the same kind of public outcry and reaction against aerosol sprays in Europe and other parts of the world as that which occurred in the US over ten years earlier. Advertisements now abound extolling the merits of ozone-friendly products. 'A good hairstyle needn't cost the Earth', is one of the latest slogans. Governments were pressured by public concern to reach agreement on the Montreal Protocol. A consumer boycott threatened by Friends of the Earth in the UK reaped dividends before it had actually started as major manufacturers and retailers, such as the supermarket chains, promised to bring alternatives onto the market at the earliest possible opportunity. While the political debate raged on, consumers were taking direct action, simply refusing to buy products that damaged the environment. And this time, consumer awareness was not restricted to CFCs and spray cans but extended into many other areas of environmental concern. There is still no legislation in the UK restricting the use of CFCs in aerosol sprays (or anywhere else) but manufacturers have largely stopped using CFCs in spray cans because they can't be sold. An 'ozone-friendly' sticker is essential if sales are to be maintained.

Prince Charles, addressing delegates at the London conference on ozone depletion in March 1989, recognized the power of consumer action: 'But that achievement [action to control CFCs] has actually been made possible by the

thousands of ordinary consumers and environmentalists whose concerned pressure persuaded the aerosol manufacturers to phase out their use of

The regional contribution to global warming during the 1980s. To date, the industrialized world is responsible for the bulk of the problem and the US is the main offender. Source: Environmental Protection Agency

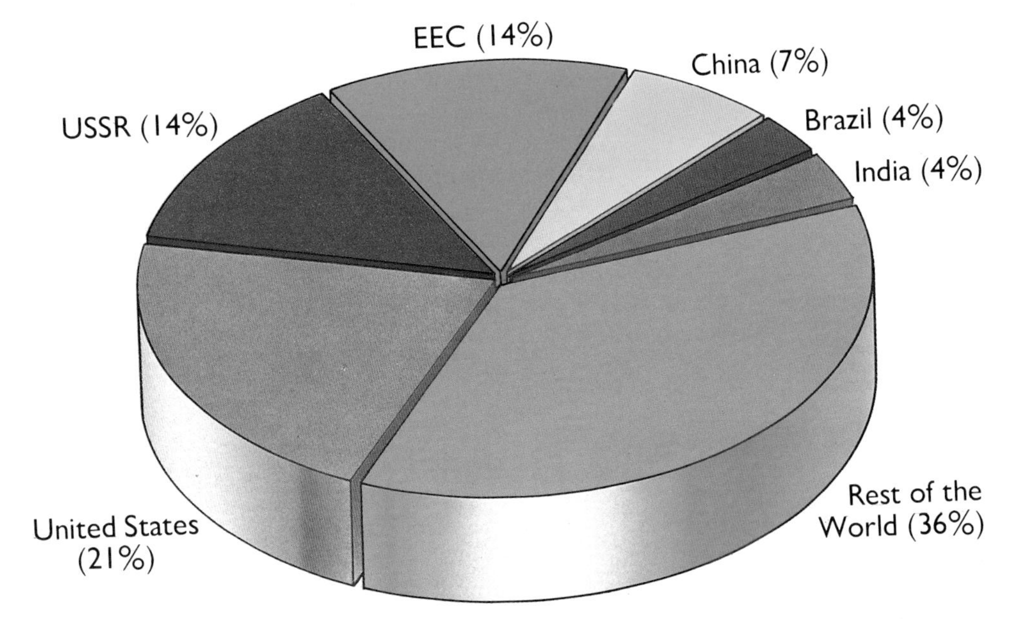

ozone-depleting CFCs by the end of the year.' He stressed that 'if we can stop the sky turning into a microwave oven, we still face the prospect of living in a garbage dump . . . Unless we realize that all these problems hang together, so will we.'

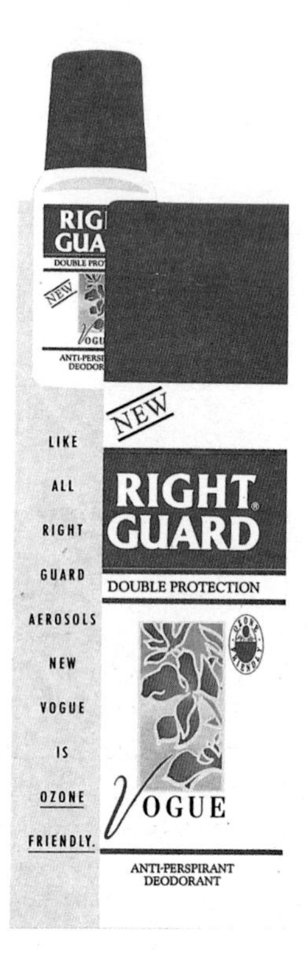

The influence of public opinion can also be seen at the political level. Even after years of negotiation, progress towards what eventually became the Montreal Protocol seemed hopelessly deadlocked in the spring of 1987. To a large extent, this was because the UK was reluctant to take action. As the European Community wanted to speak with one voice during the negotiations, British intransigence acted as a veto, removing a block of 12 votes from the international lobby in favour of a strong protocol. The situation altered dramatically in July of that year, when the UK suddenly changed its tune, swinging the whole block of European Community votes behind the move towards agreement to limit CFC releases. Interestingly, at the same time that the government changed its attitude, British industry, in the shape of ICI, a major CFC producer, also realized that the search for alternatives for CFCs would not be the lengthy process that it had previously maintained but could be achieved within a matter of a couple of years.

There were several possible reasons suggested by commentators for the change of heart by the British government. Certainly, more than one factor was at work. But, amongst the many suggestions, one of the most plausible is the fact that, in July 1987, the UK was on the verge of a general election, and the government had begun to realize that there were millions of voters who wanted something done to protect the environment in general and the ozone layer in particular. Taking a lead on environmental issues was perceived, for the first time by a major political party in the UK, as being a significant vote winner. And that perception now extends to tackling the greenhouse effect – or, at least, to stating concern!

A PLAN FOR INDIVIDUAL ACTION

Let's get down to specifics and consider precisely what combating the greenhouse effect entails. We are all responsible for this problem. We must all play a part in its solution. And there are many actions that can be taken, ranging from what you take off the supermarket shelf to lobbying government and industry directly. In fact, any action to protect the environment, even if it doesn't involve reducing emissions of the greenhouse gases directly, is important. As well as limiting other problems, it will mean that nature will be better able to withstand the onslaught of global warming. The following list of suggestions is not comprehensive but it is intended to provide a starting point. Many more good ideas can be found in books such as the *Green Consumer Guide*. If we sound as if we are preaching, please forgive us. Neither of us follows all these recommendations – but we are trying. Every little bit helps.

Recycle!

Recycling results in a range of benefits. It preserves valuable resources such as the world's forests by avoiding the unnecessary use of raw materials. It also saves energy because, in general, it takes less energy to produce a recycled product than it does to go through the whole chain of processes from raw material to finished product. And it reduces emissions from waste disposal

sites. Look at your dustbin just before the refuse is collected. Set yourself a target of, say, halving the number of sacks of garbage you produce each week. Kitchen and garden refuse should go on a compost heap. If you don't have a garden, there will probably be someone nearby who would value your contribution to their compost heap – and you might benefit from organic vegetables in return. Take glass bottles and jars to a local collection for recycling. If there isn't one, pressure your council. See if you can set up collection points for aluminium waste, for paper waste. Perhaps local businesses would sponsor such schemes. It could result in valuable publicity for the cause. Many products come shrouded by unnecessary packaging. They are then further wrapped in paper or plastic bags by the shop assistant; this additional packaging is also unnecessary. Don't buy products which are excessively packaged. Eventually manufacturers will take note. Don't be afraid to refuse unnecessary bags, paper or plastic, and use your own bag or recycled plastic bags whenever possible. Eventually, shop assistants will stop automatically packing your purchases and simply give you a receipt as proof of purchase. And the contents of your dustbin will shrink. As, in time, will your bills. Packaging costs money.

Cut down on vehicle emissions

Vehicles are a major source of a whole range of pollutants – lead, the chemicals that lead to photochemical smog and acid rain, and carbon dioxide and other greenhouse gases. Action in this area can take two forms. First, reduce emissions. If your car will run on lead-free petrol, then switch to that fuel. In many countries, it's cheaper. If you are not sure whether or not your car will run on lead-free petrol, then ask your local garage or, better still, the manufacturer. Fit a catalytic converter if that is possible. Simply asking if it can be done is an important step even if you get no further. Make the manufacturers aware that there is a demand. Then, drive more carefully. Don't put your foot down hard on the accelerator. Drive at the most fuel-efficient speed for your automobile. Pollution control devices do tend to increase the amount of gas or petrol that you use but that loss can be made up by more energy-conscious driving. But, second, and more importantly, avoid use of private vehicles wherever possible. Ask your place of work to organize a car-pool scheme. Walk, it's healthier. Use bicycles and public transport. If you don't have bicycle paths or your public transport system is inadequate, pressure your local council. Make your voice heard in support of plans to close down inner city areas to private vehicles.

Eliminate chlorofluorocarbons

The ozone wars have demonstrated beyond doubt the power of consumer action. But the battle isn't over. It is still necessary to check whether or not aerosol cans contain CFCs. Few countries have laws restricting their use and labelling isn't mandatory. Pressure the government to introduce controls and to make labelling of the contents of products a statutory requirement. Some substitutes for CFCs are powerful greenhouse gases. Make sure that claims of environmental-friendliness are true! And there are many other areas where CFCs are still used. If you buy a new refrigerator, ask the salesperson whether or not it contains CFCs. Most do; it is still difficult to find a 'green' fridge. That will change in time. Ask how efficient the fridge is in terms of energy use. Fridges are major consumers of electricity and vary greatly in their power use. Again, labelling would help consumers make the right choice. Many fast-food chains are still using packaging produced with CFCs. Ask them when they plan to phase out these products. Organize a boycott if you don't get a satisfactory response – and turn up outside the restaurant to publicize the campaign.

Reduce energy consumption

This is the most direct way of combating the greenhouse effect, lessening the use of fossil fuels and slowing down emissions of carbon dioxide. Improve the insulation in your home. Ask what can be done to reduce energy consumption in your work-place. Buy energy-efficient light bulbs – the same light for a fifth of the electricity. If they are not available in your local store, ask why not. Become more aware of how energy is used in the home. Turn down the heating a degree or two and wear extra clothing. Don't heat and light rooms that you are not using. Avoid unnecessary air-conditioning. Put lids on cooking pans. Turn electrical equipment – television, video, hi-fi – off when they are not in use. It is worthwhile spending a few minutes walking round the house, noting all the uses of energy and identifying areas where you tend to be wasteful. In many parts of the world, it is possible to use alternative sources of energy: wind power, solar power, heat pumps. If you can afford the initial investment, converting to these energy sources will reduce the greenhouse effect.

Help nature

We hamper nature's efforts to protect itself in many ways. Before helping nature directly, we need to cut out some of these practices. For example, indiscriminate use of agrochemicals – fertilizers and pesticides – is responsible for un-

told damage to the environment, harming natural ecosystems, releasing pollutants into rivers and nitrous oxide into the atmosphere. In the garden, use your compost heap: organic methods of cultivation are far preferable. This serves two purposes. It reduces pollution and environmental damage; it also leads to greater awareness of the way in which nature regulates itself, of the complex web of interactions that we brutally sever when we apply modern technology to agricultural production. In the shops, buy organic produce. If you can't, keep track of the debate on the use of harmful chemicals and ask whether or not your shopkeeper can guarantee that the latest culprit has not been used in producing your fruit or vegetables. Then, help nature to regain its ability to regulate the environment. Protect remaining areas of vegetation – don't pave over your garden – and plant trees. Tree planting is one of the most effective means of combating the greenhouse effect and there is tremendous scope for action, at the individual level and at the level of the local community. If there is waste ground in your area, suggest to your local authority that they plant trees and vegetation. Only trees that are appropriate for that habitat, of course.

Help the Third World

We tend to think that the only way to help the poor of the Third World is to give money. There's no doubt that that's a valuable response, but there are other ways. Why not tackle the problem of Third World debt? The politicians and bankers have failed but perhaps we can provide the leverage that's necessary to end this travesty. Identify the banks in your country that have lent money to the Third World. In the UK, the big four are Lloyds, Midland, Barclays and the National Westminster. If you have an account with one of them, write on the back of each cheque you use, 'I don't want my lifestyle subsidized by the poor of the Third World. Write-off Third World debt now!' The message will get through. If you don't have an account with one of these banks, open one with a small amount of money. Send messages to the bankers as above. After a while, close the account down, writing to the bank and local press explaining why you have done this. Organize a campaign and make sure the press are around when you turn up at the bank as a group to close down your accounts. Banks are very vulnerable on this issue. Unpaid debts are a great liability. Take advantage of their fear of adverse publicity. You can also help the Third World by consumer pressure in your purchasing. If you can, only buy Third World goods from outlets that you know have paid a fair price to the local producer. In fact, the best course of action would be to send a contribution to an appropriate aid agency whenever you buy any product from the Third World because, even with the best will in the world on the part of the distributor, the price the manufacturer will have been paid is unlikely to have been a fair one. Check whether or not raw materials – tropical hardwoods, for example – have been harvested sustainably. Use wood from temperate trees. Avoid buying the products of multinational companies, such as Nestlé, which have reputations for exploiting the Third World in one way or another. It may seem somewhat intangible, but redressing the balance between rich and poor is essential if global warming is to be countered.

Raise awareness

Understanding the nature of the threat is an important first step in dealing with any problem. There are more and more books about the greenhouse effect being written and documentaries shown on television. Public lectures and discussions are an excellent source of information and will enable you to meet like-minded people. Don't restrict yourself to learning about the greenhouse effect; there are many other environmental crises and the broader your knowledge, the more effective your action. Groups such as Friends of the Earth, Greenpeace, Survival International and the World Wide Fund for Nature are valuable sources of information and advice. These organizations are all supported by public donation so help them; give time – many need volunteers to help with campaigns and other activities – and money. Once you have learnt enough, then share that understanding. Give talks, to local community groups, to school, in the pub.

Let your politicians know you want action

Politicians do listen, whatever the nature of the political system. You just have to speak loudly. Writing letters to your local representative places the issue on his or her agenda. It's often more effective to ask a specific question rather than to simply give your opinion – a question which will lead the politician to ask a particular government department or minister for their response to your query. That way, you reach more than one person. What is the Department of Agriculture doing about the threat to production on coastal areas posed by sea level rise? How much money is the Energy Minister committing to improving energy efficiency over the coming year? How can high levels of military spending be justified, at the expense of environmental protection, when global security is threatened by climatic change? If you go through your local politician, it is more likely that a departmental response will be forthcoming. Don't neglect community politicians – your city or rural council. When election time comes round, go to public meetings and tackle the candidates directly on the environmental stance. They are more likely to commit themselves at this point; you can then remind them of their words after they get elected.

Stop overconsuming

As we have argued throughout this book, overconsumption in the industrialized world is one of the prime causes of global warming. And we are all responsible for this crime. We are using more than our fair share of the planet's resources. People are already suffering as a result, and that suffering will increase as climate changes and sea level rises. Let's start with food. Cut down on meat consumption. Not only is meat a very inefficient means of using the nutrition in grain but it is North American consumption of beef which is a driving force behind the destruction of the tropical rainforest. South American forests are destroyed to make way for grassland to feed the beef that goes into North American hamburgers. Avoid unnecessary purchases. Energy has been used in the production of everything that you buy. If you buy less, you save energy and combat the greenhouse effect. Don't replace last year's model with this year's simply because the adverts say you should. Fix broken goods wherever possible, rather than immediately going out to the shops to buy a replacement. Think about what it is that makes your life worthwhile. Is it spending money on new goods, or is there more to life than consumerism? Inundated by advertising, it is difficult to resist the feeling that a purchase will make you feel better, will make the world appear a better place. Is this really true?

A CHALLENGE AND AN OPPORTUNITY

Global warming is not an isolated problem. It is just one of many interlinked threats to the environment and to the future of life on Earth. Perhaps the most significant aspect of the greenhouse effect is that it has highlighted a whole suite of problems and has provided a much-needed focus for action. It has brought to the fore that fact that, had we treated the planetary environment – and our fellow citizens – with greater respect, we would not be facing crisis today.

In the words of the World Commission on Environment and Development in *Our Common Future*: 'The next few decades are crucial. The time has come to break out of past patterns. Attempts to maintain social and ecological stability through old approaches to development and environmental protection will increase instability. Security must be sought through change.'

It is a striking aspect of the means by which we might limit the global warming that many of the options make good sense anyway. We can reduce heating bills, extend the life of our reserves of fossil fuels, minimize the problems of acid rain and ozone depletion, protect the tropical rainforests, improve global security and the economic health of the Third World. And by taking action to tackle global warming, we will also reduce poverty, inequality and oppression. It is a sad comment on the state of our society that it may take the threat of catastrophe to force such change.

This is the challenge that the global warming presents – a challenge, and an opportunity to create a more just, equitable and safer society. What value is there in modern civilization – what have we gained – if the quality of life that has been developed is not available for all?

FURTHER READING

For more detailed information, at about the same level as this book, on the scientific background to the greenhouse effect, the history of its investigation and the way in which the scenarios for the twenty-first century are derived, see *Hothouse Earth* by John Gribbin (Bantam Press, 1990).

Several of John Gribbin's books deal with topics related to those discussed here. *Future Worlds* (Plenum, 1981) looks at the problems of inequality and the paths to a more equitable society, drawing on research carried out by the Science Policy Research Unit, University of Sussex; *Children of the Ice* (with Mary Gribbin, Blackwell, 1990) describes how past climatic changes have influenced humankind, and puts the greenhouse effect in a longer-term perspective; and *The Hole in the Sky* (Bantam/Corgi, 1988) looks at the specific problem of the depletion of the ozone layer caused by CFCs, which are also greenhouse gases.

Friends of the Earth (UK) have published *The Heat Trap* (1988), a briefing document co-authored by Mick Kelly. This covers the scientific background to the problem and details appropriate policy responses. It contains references to key sources of information.

The definitive technical discussion of the heat trap is *The Greenhouse Effect, Climatic Change and Ecosystems*, edited by Bert Bolin, Bo Döös, Jill Jäger and Richard Warrick for the Scientific Committee on Problems of the Environment (Wiley, 1986; this is the report known as 'SCOPE 29'). And the best up-to-date assessment of the problems facing human society as we move into the twenty-first century, and how to tackle them, is the report *Our Common Future*, from the World Commission on Environment and Development (Oxford University Press, 1987; sometimes referred to as the Brundtland Report). For further suggestions for individual action, see *The Green Consumer Guide* by John Elkington and Julia Hailes (Gollancz, 1988).

Other sources of information that we have used include: *Energy for a Sustainable World* by Jose Goldemberg, Thomas Johansson, Amulya Reddy and Robert Williams (Wiley, 1988); *Developed to Death* by Ted Trainer (Green Print/Merlin Press, 1989); *A Fate Worse than Debt* by Susan George (Penguin, 1988); *Africa in Crisis* by Lloyd Timberlake (Earthscan, 1985); *Earth* by Anne and Paul Ehrlich (Thames/Methuen, 1987); and *Gaia: An Atlas of Planet Management* edited by Norman Myers (Anchor Press/Doubleday, 1984). The annual publication *State of the World* (Norton) prepared by the Worldwatch Institute, Washington DC, is an excellent source of up-to-date information. More frequent updates can be found in the periodicals *New Scientist* and *The Ecologist*.

The Toronto Conference, *The Changing Atmosphere: Implications for Global Security*, crystallized the growing recognition that the critical state of the planetary environment warrants prompt and decisive action. Citing the impact of global warming as 'second only to nuclear war', the delegates recommended that immediate action should be taken to:

- reduce carbon dioxide emissions by 20% of 1988 levels by the year 2005 through improved energy efficiency and modification of supply;
- halt deforestation and increase afforestation;
- strengthen the 1987 Montreal Protocol concerning control of chlorofluorocarbon emissions in order to eliminate releases by the year 2000, and to reduce emissions of other greenhouse gases;
- initiate the development of a comprehensive global convention for protocols on the protection of the atmosphere;
- devote increased resources to research programmes concerned with scientific and policy aspects of the problems, to support the continuing assessment of research results and to stimulate governmental discussion of responses and strategies;
- establish a trust fund to assist Third World nations and to encourage these nations to participate in international efforts concerning monitoring, research, adaptation and control; and
- increase funding to non-governmental organizations and educational establishments to permit the establishment and development of educational campaigns and programmes.